T0137224

Studies in Computational Intelligence

Volume 787

Series editor

Janusz Kacprzyk, Polish Academy of Sciences, Warsaw, Poland
e-mail: kacprzyk@ibspan.waw.pl

The series "Studies in Computational Intelligence" (SCI) publishes new developments and advances in the various areas of computational intelligence—quickly and with a high quality. The intent is to cover the theory, applications, and design methods of computational intelligence, as embedded in the fields of engineering, computer science, physics and life sciences, as well as the methodologies behind them. The series contains monographs, lecture notes and edited volumes in computational intelligence spanning the areas of neural networks, connectionist systems, genetic algorithms, evolutionary computation, artificial intelligence, cellular automata, self-organizing systems, soft computing, fuzzy systems, and hybrid intelligent systems. Of particular value to both the contributors and the readership are the short publication timeframe and the world-wide distribution, which enable both wide and rapid dissemination of research output.

More information about this series at http://www.springer.com/series/7092

Roger Lee

Editor

Computational Science/Intelligence & Applied Informatics

 Springer

Editor
Roger Lee
Software Engineering and Information
  Technology Institute
Central Michigan University
Mount Pleasant, MI, USA

ISSN 1860-949X   ISSN 1860-9503   (electronic)
Studies in Computational Intelligence
ISBN 978-3-030-07255-1   ISBN 978-3-319-96806-3   (eBook)
https://doi.org/10.1007/978-3-319-96806-3

This Springer imprint is published by the registered company Springer Nature Switzerland AG
The registered company address is: Gewerbestrasse 11, 6330 Cham, Switzerland

Foreword

The purpose of the 5th ACIS International Conference on Computational Science/Intelligence & Applied Informatics (CSII 2018) which was held on July 10–12, 2018 in Yonago, Japan was to together researchers, scientists, engineers, industry practitioners, and students to discuss, encourage, and exchange new ideas, research results, and experiences on all aspects of A Computational Science/ Intelligence & Applied Informatics and to discuss the practical challenges encountered along the way and the solutions adopted to solve them. The conference organizers have selected the best 13 papers from those papers accepted for presentation at the conference in order to publish them in this volume. The papers were chosen based on review scores submitted by members of the program committee and underwent further rigorous rounds of review.

In Chapter "GUI Testing for Introductory Object-Oriented Programming Exercises", Ushio Inoue presents a method to test and score student programs with graphical user interfaces written in JavaFX. The method is based on scripts that analyzes the structure of programs under test and simulates user's interactions.

In Chapter "Python Deserialization Denial of Services Attacks and Their Mitigations", Kousei Tanaka and Taiichi Saito precisely describe the DoS attacks and their mitigations using a specially crafted data that consumes huge memory in deserialization. There is a possibility that the memory consumption leads to denial of services attacks.

In Chapter "A Branch-and-Bound Based Exact Algorithm for the Maximum Edge-Weight Clique Problem", Satoshi Shimizu, Kazuaki Yamaguchi, and Sumio Masuda propose an exact algorithm based on branch-and-bound. By some computational experiments, they confirmed that their proposal algorithm is faster than the methods based on mathematical programming.

In Chapter "A Software Model for Precision Agriculture Framework Based on Smart Farming System and Application of IoT Gateway", Symphorien Karl Yoki Donzia, Haeng-Kon Kim, and Ha Jin Hwang propose a framework for precision agriculture using IoT for solving human food, and the productivity of crops must be increased first. IoT solution through architecture, platforms, and IoT standards, or the use of interoperable IoT technologies beyond the adopters, in particular, simplifies the existing proposals.

In Chapter "Components of Mobile Integration in Social Business and E-commerce Application", Mechelle Grace Zaragoza, Haeng-Kon Kim, and Youn Ky Chung proposed an application of mobile integration components in social enterprise and e-commerce as a software development methodology to simply integrate different basic components of the technology into a single web-based solution. They propose a systematic development process for the software agent using components and UML.

In Chapter "Design and Evaluation of a MMO Game Server", Youngsik Kim, Ki-Nam Kim implement a simple MMO game server using IOCP and evaluate its performance. The simple MMO game server implemented in this paper also supports multi-thread synchronization and dead reckoning.

In Chapter "A Study on the Characteristics of Electroencephalography (EEG) by Listening Location of OLED Flat TV Speaker", Hyungwoo Park, Sangbum Park, and Myung-Jin Bae have analyzed the acoustic characteristics of a directly driving the OLED panel speakers, using electroencephalography (EEG); in addition, they study the advantages of direct driving sound.

In Chapter "A Study on Design of Efficient Private Blockchain", Ki Ho Kwak, Jun Taek Kong, Sung In Cho, Huy Tung Phuongand Gwang Yong Gim present an issue of a reliable private blockchain that will be derived and design measures be proposed to make a contribution so as to enable safe utilization in the environment of P2P distributed network.

In Chapter "Designing System Model to Detect Malfunction of Gas Sensor in Laboratory Environment", Ki-Su Yoon, Seoung-Hyeon Lee, Jae-Pil Lee, and Jae-Kwang Lee collect correlation data of temperature sensor and gas sensor to prevent this. After confirming the correlation through correlation analysis, they calculate regression coefficient by regression analysis and obtain regression equation that can extract sample values of gas sensor data and temperature sensor data.

In Chapter "Design of Device Mutual Authentication Protocol in Smart Home Environment", Jae-Pil Lee, Seoung-Hyeon Lee, Jae-Gwang Lee, and Jae-Kwang Lee suggest a security protocol for device/user authentication and access control in order to enable easy and convenient compatibility services. In addition, it gives you a way to establish a safe and secure framework through the protocol suggested.

In Chapter "A Study of Groupthink in Online Community", Nhu-Quynh Phan, Seok-Hee Lee, Jin Won Jang, and Gwang Yong Gim present a study about groupthink phenomenon that can affect the decision-making quality of online communities in Korea. This empirical study toward the respondents was in Korea who currently participate online communities.

In Chapter "Development of a Physical Security Gateway for Connectivity to Smart Media in a Hyper-Connected Environment", Yong-Kyun Kim, Geon Woo Kim, and Seoung-Hyeon Lee propose a method of blocking the intrusion from the external through unidirectional communication method, to help prevent the risk of leakage of personal information. There is a great need for a method of protecting the personal data of the server from the invasion of the server.

In Chapter "Design and Implementation of Security Threat Detection and Access Control System for Connected Car", Joongyong Choi, Hyeokchan Kwon, Seokjun Lee, Byungho Chung, and Seong-il Jin design a whitelist-based access control system to detect and block malicious attempts to access an in-vehicle network through an infotainment device in a connected car environment and present the implementation results.

It is our sincere hope that this volume provides stimulation and inspiration, and that it will be used as a foundation for works to come.

Tokyo, Japan   Takayuki Fujimoto
July 2018   Toyo University

Contents

Contributors

Myung-Jin Bae School of Information and Technology, Soongsil University, Seoul, Republic of Korea

Sung In Cho Department of IT Policy and Management, Department of Business Administration, Soongsil University, Seoul, South Korea

Joongyong Choi Information Security Research Division, Electronics and Telecommunications Research Institute, Daejeon, South Korea

Byungho Chung Information Security Research Division, Electronics and Telecommunications Research Institute, Daejeon, South Korea

Youn Ky Chung Kyungil University, Daegu, Korea

Symphorien Karl Yoki Donzia Daegu Catholic University, Gyeongsan, South Korea

Gwang Yong Gim Department of Business Administration, Soongsil University, Seoul, South Korea

Ha Jin Hwang Sunway University Business School, Sunway University, Subang Jaya, Malaysia

Ushio Inoue Department of Communication and Information Engineering, Tokyo Denki University, Tokyo, Japan

Jin Won Jang Business Administration, Soongsil University, Seoul, Korea

Seong-il Jin Chungnam National University, Daejeon, South Korea

Geon Woo Kim Information Security Research Division, Electronics and Telecommunications Research Institute, Daejeon, Korea

Haeng-Kon Kim Daegu Catholic University, Gyeongsan, South Korea; Daegu Catholic University, Daegu, South Korea

Ki-Nam Kim Department of Game and Multimedia Engineering, Korea Polytechnic University, Siheung, Republic of Korea

Yong-Kyun Kim Information Security Research Division, Electronics and Telecommunications Research Institute, Daejeon, Korea

Youngsik Kim Department of Game and Multimedia Engineering, Korea Polytechnic University, Siheung, Republic of Korea

Jun Taek Kong Department of Business Administration, Soongsil University, Seoul, South Korea

Ki Ho Kwak Department of Business Administration, Soongsil University, Seoul, South Korea

Hyeokchan Kwon Information Security Research Division, Electronics and Telecommunications Research Institute, Daejeon, South Korea

Jae-Gwang Lee Department of Computer Engineering, Hannam University, Daejeon, South Korea

Jae-Kwang Lee Department of Computer Engineering, Hannam University, Daejeon, South Korea

Jae-Pil Lee Department of Computer Engineering, Hannam University, Daejeon, South Korea

Seok-Hee Lee Information Technology Policy Management, Seoul, Korea

Seokjun Lee Information Security Research Division, Electronics and Telecommunications Research Institute, Daejeon, South Korea

Seoung-Hyeon Lee Information Security Research Division, Electronics and Telecommunications Research Institute, Daejeon, South Korea

Sumio Masuda Kobe University Faculty of Engineering, Nada, Kobe, Japan

Hyungwoo Park School of Information and Technology, Soongsil University, Seoul, Republic of Korea

Sangbum Park School of Information and Technology, Soongsil University, Seoul, Republic of Korea

Nhu-Quynh Phan Business Administration, Soongsil University, Seoul, Korea

Huy Tung Phuong Department of Business Administration, Soongsil University, Seoul, South Korea

Taiichi Saito Tokyo Denki University, Graduate School of Engineering, Tokyo, Adachi-Ku, Japan

Satoshi Shimizu Kobe University Faculty of Engineering, Nada, Kobe, Japan

Kousei Tanaka Tokyo Denki University, Graduate School of Engineering, Tokyo, Adachi-Ku, Japan

Kazuaki Yamaguchi Kobe University Faculty of Engineering, Nada, Kobe, Japan

Ki-Su Yoon Hannam University Daejeon, Daejeon, Korea

Mechelle Grace Zaragoza Daegu Catholic University, Daegu, South Korea

GUI Testing for Introductory Object-Oriented Programming Exercises

Ushio Inoue

Abstract Automated testing is necessary in large classrooms where many students learn a programming language. This paper presents a method to test and score student programs with graphical user interfaces written in JavaFX. The method is based on scripts that analyzes the structure of programs under test and simulates user's interactions. We implemented several utility methods to write the testing scripts easy. No additional software library is required to run the scripts. Preliminary evaluation results are shown on the developing and executing of scripts for real exercises in our introductory programming classrooms.

Keywords GUI testing · Programming education · Automated scoring · Java application · JavaFX

1 Introduction

Many university students learn GUI (Graphical User Interface) programming. GUI programming is fun for beginners. It also helps to understand the important concepts of OOP (Object Oriented Programming) such as encapsulation, inheritance, polymorphism, and dynamic binding.

Practice is very important in learning any programming language. Therefore, teachers demand students in their classrooms to write many programs. These programs written by students have to be tested and scored. An obvious way to test such programs is manual testing, in which the teachers interact with the GUI directly by their own hand. However, a large classroom causes a problem to the teacher. The teacher has to check many programs in a short period of time and score them on a unified criterion. Manual testing is not a practical solution. Thus, automated testing is necessary in large classrooms.

U. Inoue (✉)
Department of Communication and Information Engineering,
Tokyo Denki University, Tokyo, Japan
e-mail: inoueu@mail.dendai.ac.jp

© Springer International Publishing AG, part of Springer Nature 2019
R. Lee (ed.), *Computational Science/Intelligence & Applied Informatics*, Studies in Computational Intelligence 787, https://doi.org/10.1007/978-3-319-96806-3_1

Automated GUI testing is not easy for several reasons [1]: First, GUIs are designed for humans not computer programs. Second, conventional unit-testing is not suitable because a single event may affect more than one GUI components. Third, user-generated events must be simulated to interact GUI components. Fourth, slight differences in the layout of GUI components should not affect the test results.

Three common approaches for automated GUI testing are script-based, model-based, and CR-based [2]. In the first approach, the user (the teacher in the classroom) writes a script that interacts GUI components programmatically. This task is troublesome and error-prone if the target GUI is complicated. In the second approach, the user specifies a test model in a formal modeling language such as the UML (Unified Modeling Language) [3]. The test model is used for automated test case generation. This approach requires a complicated system that interprets the model. The third approach is based on a Capture/Replay or Record/Playback technique [4]. The user runs the GUI and records the entire interactive session in a log file. Given that file, the user can then automatically replay the exact same interactive session without requiring manual operations. This approach is not suitable to test many similar but different programs.

This paper describes a GUI testing method for introductory OOP exercises. The method is script-based, because each exercise is designed to create a relatively simple GUI application using the JavaFX framework [5]. It is not difficult to write scripts to test them. Rather, it is important to quickly test many similar but different programs that students submitted. Another merit of the script-based approach is that it is independent of the development environment used by the students. Each student can use any environment to write programs and test them without installing any additional software library.

The rest of the paper is organized as follows: Sect. 2 describes related work. Section 3 gives an overview of JavaFX. Section 4 explains the prerequisites on our exercises, and Sect. 5 presents the proposed method and its implementation. Section 6 shows the evaluation, and Sect. 7 concludes the discussion.

2   Related Work

Automated testing of student programs has been an important research topic. Ihantola et al. [6] surveyed the developments of automatic assessment tools for programming exercises by systematically collecting research papers published between 2006 and 2010.

In recent years, research and development of automated testing for GUI applications written in Java Swing/AWT has been very active. Gray and Higgins [7] presented an introspective approach for the marking of GUI. They use the knowledge of the hierarchical structure of OOP to access their internal components. Thornton et al. [8] supported students writing self-tests of GUI programs to practice TDD (Test Driven Development). To support students in introductory level, they adapted the Objectdraw student graphic library. Ahmadzadeh et al. [9] proposed an Eclipse plugin named JavaMarker. The plugin uses test cases written in a plain text file. Their system seems to be limited to testing very simple GUIs. Snyder et al. [10] developed a library for interface testing: LIFT, which supports students to write GUI tests for Java applications. LIFT is script-based and an extension of Objectdraw. Our method is different from LIFT in that it requires nothing except JavaFX.

Few research papers have been published on automated testing of JavaFX applications. Kruk et al. [11] studied three approaches to automated GUI testing, and implemented a unit test runner extending the standard JUnit runner for JavaFX. Klammer et al. [12] developed a test case generator for unit testing and system testing. Their generator is based on the TestFX framework, which allows writing JUnit scripts for JavaFX GUI controls and applications.

3 JavaFX

JavaFX [5] is a new GUI framework for Java applications designed as a replacement of Swing/AWT, which has been in wide spread use for a long time. JavaFX is written in pure Java and easy to learn for Java programmers. It supports modern GUI control functions that can easily incorporate special effects. Based on the feature of the scene graph, JavaFX makes it simple to manage the window of the program. JavaFX also supports the use of CSS, FXML, and Scene Builder to style and build a GUI. The current version is JavaFX 8, which is included with the standard JDK and JRE bundles. This section explains the scene graph, which is an important concept of JavaFX and the basis of learning OOP.

The JavaFX framework is provided by dozens of packages consisting of many Java classes. The main class for a JavaFX application is a subclass of the Application class. The start method is the main entry point. A JavaFX application defines the user interface containers. The top-level container is the Stage class that displays a window on the display device. It contains the Scene class, and the content of the scene is represented as a hierarchical scene graph of nodes. Figure 1 shows a simple scene graph example. In this graph, the root node is the Pane class, which manages the layout of nodes inside it. The pane contains three control nodes: Button, Label, and TextField classes.

Here is a GUI application that explains how to use JavaFX and how to test the scene graph. Figure 2 shows a screenshot of the application program checking whether the student ID typed in the textfield is valid. Figure 3 shows the JavaFX code of the program.

As shown in Fig. 3, the GUI application is created in a bottom-up fashion. First, four control nodes (two labels, a textfield and a button) are created. Then, these nodes are added in the HBox pane node, which lays out its child nodes in a single horizontal row. Then, the pane node is set in the scene. Finally, the scene is set in the primary stage and shown on the screen.

Conversely, the application can be tested in a top-down fashion. The first clue is the primary stage. The tester creates an instance of the Application class and invokes

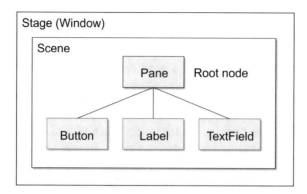

Fig. 1 Simple scene graph example

Fig. 2 Screenshot of the
GUI application program

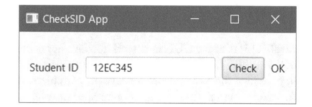

the start method, which returns the reference of the primary stage. Then, the tester
gets the scene and the HBox pane. Then, it gets children of the pane, and finds the
four control nodes. Then, it sets a student ID prepared for the test in the textfield,
and clicks on the button. Finally, it gets and verifies the value of the second label.

4 Perquisites on Exercises

In our classroom, the teacher requests each student to write a GUI application in an
exercise. The exercise includes detailed specifications on the structure and function
of the application. Figure 4 shows an exercise example that expects a student to write
the GUI application shown in Fig. 3. It specifies both the structure and function of the
GUI. The structure includes the types of components to use, the values to assign, and
the layout to apply. The function includes the student ID format and the component
behavior.

```
// Omitted import and package declarations

public class CheckSIDApp extends Application {

    @Override
    public void start(Stage primaryStage) {
        Label label1 = new Label("Student ID");
        Label label2 = new Label();
        TextField textField = new TextField();
        Button button = new Button("Check");
        button.setOnAction((ActionEvent event) -> {
            String sid = textField.getText().trim();
            if (checkSID(sid))
                label2.setText("OK");
            else
                label2.setText("NG");
        });
        HBox pane = new HBox(10, label1, textField, button, label2);
        pane.setAlignment(Pos.CENTER);
        Scene scene = new Scene(pane, 400, 100);
        primaryStage.setScene(scene);
        primaryStage.setTitle("CheckSID App");
        primaryStage.show();
    }

    public static boolean checkSID(String sid) {
        String regex = "^[0-9]{2}[Ee][Cc][0-9]{3}$";
        return sid.matches(regex);
    }
}
```

Fig. 3 JavaFX code of the application program

Fig. 4 Exercise example

> Ex1
>
> Create a GUI application that validate a student ID satisfying the following requirements:
>
> - The application is a public class named CheckSIDApp that extends the Application class.
> - It uses the HBox to layout Label, TextField, Button, and another Label in this order.
> - The user of the GUI types a student ID in the TextField, and clicks on the Button.
> - If the student ID is valid, "OK" will be displayed on the second Label. Otherwise, "NG" will be displayed on the Label.
> - A valid student ID consists of seven characters, starting with a two-digit number followed by "EC" followed by a three-digit number. The string "EC" is case-insensitive.

5 Proposed Method and Implementation

As mentioned before, we use a script-based approach to test GUI application written by students. This section defines the application under test (AUT) and describes the method used by the testing script (tester).

5.1 AUT and Tester

Each student writes an AUT in the exercise and the teacher writes a tester as shown in Fig. 5. The student will manually test the GUI created by the AUT. After the manual test is completed, the student uses the tester to verify the GUI. The tester automatically starts up the AUT and tests the GUI. The test results are recorded in the log, which are output to the console. The tester is also used by the teacher to test all AUTs written in the classroom in a batch process.

5.2 Testing Method

The tester starts the AUT and tests the GUI in three steps, which are startup, structure, and action tests.

In the startup test, the tester creates an instance of the AUT, then invokes the start method to get the scene graph in the primary stage. We wrote a utility method shown in Table 1 to get the scene currently displayed on the stage. Figure 6 shows the code of the method.

In the structure test, the tester checks the structure of the scene. The check includes the type and properties of the layout pane and the control nodes contained in the pane. We implemented methods shown in Table 2 to find the root pane and its children nodes easily. Note that the findPane and findNode methods are overloaded by having different parameters specifying how to find the target. Figures 7 and 8 show the code of such methods.

In the action test, the tester checks the actions of the nodes by simulating user's interactions. A typical test case is: setting a value in the textfield, clicking on the button, and getting the result from the label. We implemented several methods for automating the test of various action combinations. Table 3 shows some of the frequently used methods. The actTest method is also overloaded. Figure 9 shows the code of such a method.

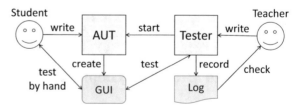

Fig. 5 System overview

Table 1 Utility method for startup test

Utility method	Description
Scene findScene(Stage stage)	Returns the current scene on the stage

```
public static Scene findScene(Stage stage) {
    System.out.print("--- Starting Scene ... ");
    Scene scene = stage.getScene();
    if (scene != null) {
        System.out.println("Success");
        return scene;
    } else {
        System.out.println("Failure");
        return null;
    }
}
```

Fig. 6 Finding the current scene on the stage

Table 2 Utility methods for structure test (partial list)

Utility method	Description
Pane findPane(Scene scene, String type)	Returns the root pane node in the scene
Pane findPane(Pane pane, String type)	Returns the first child nested pane whose pane type matches the type in the pane
Node findNode(Pane pane, String type, int n)	Returns the n-th child node in the pane, whose node type matches the type
Node findNode(Pane pane, String type, String value)	Returns the first child node in the pane, whose node type matches the type and whose text contains the value

```
public static Pane findPane(Scene scene, String type) {
    System.out.printf("--- Checking %s ... ", type);
    Parent root = scene.getRoot();
    String nodeType = null;
    if (root != null) {
        nodeType = root.getClass().getSimpleName();
        if (type.equals(nodeType)) {
            System.out.println("Success");
            return (Pane) root;
        }
    }
    System.out.printf("Failure because %s ?\n", nodeType);
    return null;
}
```

Fig. 7 Finding the root pane node in the scene

```
public static Node findNode(Pane pane, String type, int n) {
    System.out.printf("--- Checking %s ... ", type);
    ObservableList<Node> ol = pane.getChildren();
    Node node;
    String nodeType = null;
    if (ol != null && n < ol.size()) {
        node = ol.get(n);
        nodeType = node.getClass().getSimpleName();
        if (type.equals(nodeType)) {
            System.out.println("Success");
            return node;
        }
    }
    System.out.printf("Failure because %s ?\n", nodeType);
    return null;
}
```

Fig. 8 Finding the n-th child node in the pane

Table 3 Utility methods for structure test (partial list)

Utility method	Description
Boolean actTest(TextField tf, Button bt, Label lb, String vi, String vo)	Sets the vi in the TextField, then clicks on the Button, and gets the result from the Label. Returns true if the result matches the vo
Boolean actTest(TextField tf1, Button bt, TextField tf2, String vi, String vo)	Sets the vi in the TextField tf1, then clicks on the Button, and gets the result from the tf2. Returns true if the result matches the vo
Boolean actTest(Button bt, Label lb, String vo)	Clicks on the Button, and gets the result from the Label. Returns true if the result matches the vo

```
public static boolean actTest(TextField tf, Button bt, Label lb,
        String in, String out) {
    System.out.printf("--- Text data \"%s\" ... ", in);
    tf.setText(in);
    bt.fire();
    String res = lb.getText();
    if (res.contains(out)) {
        System.out.println("Success");
        return true;
    }
    System.out.printf("Failure because %s ?\n", res);
    return false;
}
```

Fig. 9 Checking the behavior of the button

5.3 Scoring Method

The scoring policy varies depending on the teacher and exercise. Therefore, a simple and flexible scoring method is implemented based on our previous work [13]. The score of each AUT is determined with the following linear formula:

$$Score = \frac{S_1 \times N_1 + S_2 \times N_2 + S_3 \times N_3}{S_{\max}} \times 100 \, (\%)$$

where S_i is the base score in the i-th step, N_i is the number of nodes or test cases successfully passed in the step, and S_{max} is the max score. The tester counts the number of successes during the execution of each step.

Figure 10 shows the tester code for the GUI application shown in Fig. 3. First, the tester opens the stage and gets the scene. Then, it gets the HBox pane and gets two labels, a text field, and a button. Then, it types a student ID in the text field, clicks on the button, and validates the result shown on the second label. This is repeated four times with different student IDs. Finally, the tester close the stage.

5.4 Limitations

The proposed method has the following restrictions, but they rarely cause problems in introduction exercises:

- Only specified components must be used.
  Our method finds the GUI components with the standard JavaFX methods. The GUI components created by AUT must be exactly as specified in the exercise. If an unexpected component is found, the test will fail.
- Public member variables are needed.
  Our method first gets the primary stage, then finds the components. If there is something that cannot be found in this way, such as a secondary stage or an alert box, its reference must be defined as a member variable outside methods.
- Manual response is needed on an alert box.
  According to the specification of JavaFX, when displaying the alert box, the program waits for user's manual response. Therefore, fully automatic testing is not possible if alert boxes are used.
- No concurrency is allowed.
  The scene graph is not thread-safe and can only be accessed and modified from the JavaFX application thread. If a new thread is created and executed in parallel, it may result in incorrect test results.

```
public class CheckSIDAppTester extends Application {

    CheckSIDApp app;
    String appName = "CheckSIDApp";
    int score = 0, scoreMax = 10, s1 = 1, s2 = 1, s3 = 1;

    @Override
    public void start(Stage primaryStage) {
      try { System.out.printf("Start to test %s\n", appName);
        app = new CheckSIDApp();
        app.start(primaryStage);

        Scene scene = findScene(primaryStage);
        if (scene != null)
            score += s1;
        else {
            printScore();
            return;
        }

        HBox pane = (HBox) findPane(scene, "HBox");
        if (pane != null)
            score += s2;
        else {
            printScore();
            return;
        }

        Label lb1 = (Label) findNode(pane, "Label", 0);
        if (lb1 != null)
            score += s2;
        TextField tf1 = (TextField) findNode(pane, "TextField", 1);
        if (tf1 != null)
            score += s2;
        Button bt1 = (Button) findNode(pane, "Button", 2);
        if (bt1 != null)
            score += s2;
        Label lb2 = (Label) findNode(pane, "Label", 3);
        if (lb2 != null)
            score += s2;
        if (tf1 == null || bt1 == null || lb2 == null) {
            printScore();
            return;
        }

        String[][] tcs = {
            {"12EC34", "NG"}, {"23EC456", "OK"}, {"34EC5678", "NG"},
            {"", "NG"},};
        Boolean test = true;
        for (String[] tc : tcs) {
            test = actTest(tf1, bt1, lb2, tc[0], tc[1]);
            if (test)
                score += s3;
            else {
                printScore();
                return;
            }
        }
        primaryStage.close();
        System.out.printf("Finish to test %s\n", appName);
        printScore();
      } catch (Exception e) {
        printScore();
        e.printStackTrace();
      }
    }
}
```

Fig. 10 Tester code of the application shown in Fig. 3

6 Evaluation

At the time of writing this paper, no formal evaluation has been conducted, but preliminary evaluation on developing and executing testers was done for real programming exercises in our classroom.

6.1 Developing Testers

Developing a script-based tester is tedious and error-prone in general. We evaluated the effort to write testers for GUI applications, Ex1–Ex4, shown in Fig. 11. The effort was measured by the source code size, which is the number of lines of the tester. The result is shown in Table 4. Each row in the table shows the brief description of the application, the number of test cases, and the source code size of the tester. For example, Ex1 tester that tests the application shown in Fig. 3 was written in 66 lines of code to test the 10 test cases in total. Note that the source code does not include the code of utility methods described in Sect. 5. As the number of test cases especially in the startup and structure tests increases, the code size increases. However, it is not difficult to create a new tester, because the pattern of code in each tester is very similar and reusable.

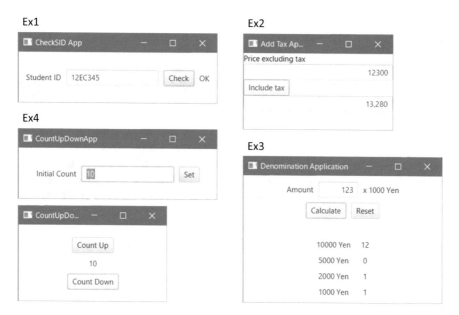

Fig. 11 Screenshots of GUI applications evaluated

Table 4 Testers used for the evaluation and their code size

Tester	Brief description	Startup test cases	Structure test cases	Action test cases	Total test cases	Code size (in lines)
Ex1	Check student ID	1	5	4	10	66
Ex2	Add tax to price	1	5	9	15	68
Ex3	Calculate denomination	1	16	8	25	111
Ex4	Count up and down	2	10	8	20	152

6.2 Executing Testers

Testing speed is another important issue in large classrooms where the teacher tests many student programs. We evaluated the average time required to compile and run the tester and AUT using a PC with Intel Core i7-7500U CPU @ 2.7 GHz. The result is shown in Table 5. For example, it requires 2.468 s in total to test a student program of Ex1. Even if the number of test cases increases, the increase of the total time is negligible. It is possible to test 100 student programs within 5 min.

7 Conclusion

This paper has presented a method to test GUI applications written in JavaFX 8 for introductory OOP exercises. Our script-based method analyzes the structure of the GUI components in a top-down fashion and verifies the behavior of those components automatically. Students can use any developing environment to write their programs without installing additional software other than the standard JDK bundle. Several utility methods have been implemented for teachers to efficiently write testers. The preliminary evaluation results on the effort to write testers and the time required to run testers show that our method is applicable to large introductory OOP classrooms.

Table 5 Average time required for testing a student program (in seconds)

Tester	Compile AUT	Compile tester	Run tester	Total
Ex1	0.610	0.636	1.222	2.468
Ex2	0.624	0.631	1.235	2.490
Ex3	0.643	0.644	1.306	2.593
Ex4	0.635	0.652	1.302	2.589

We have started using our method in a real classroom where about 120 students learn OOP and getting a good reputation. We are also working to add new features and remove restrictions to respond to requests from the teacher and students. We will report formal evaluation results in near future.

References

1. Ruiz, A., Price, Y.W.: Test-driven GUI development with TestNG and Abbot. IEEE Softw. **24**(3), 51–57 (2007)
2. Nguyen, B.N., Robbins, B., Banerjee, I., Memon, A.: GUITAR: an innovative tool for automated testing of GUI-driven software. Autom. Softw. Eng. (Springer) **21**(1), 65–105 (2014)
3. Mlynarski, M., Güldali, B., Weißleder, S., Engels, G.: Model-based testing: achievements and future challenges. Adv. Comput. (Elsevier) **86**, 1–39 (2012)
4. Adamoli, A., Zaparanuks, D., Jovic, M.: Automated GUI performance testing. Softw. Qual. J. (Springer) **19**(4), 801–839 (2011)
5. Schildt, H.: Introducing JavaFX 8 Programming. McGraw-Hill (2015)
6. Ihantola, P., Ahoniemi, T., Karavirta, V., Seppälä, O.: Review of recent systems for automatic assessment of programming assignments. In: Proceedings of the 10th Koli Calling International Conference on Computing Education Research, pp. 86–93 (2010)
7. Gray, G.R., Higgins, C.A.: An introspective approach to marking graphical user interfaces. ACM SIGCSE Bull. **38**(3), 43–47 (2006)
8. Thornton, M., Edwards, S.H., Tan, R.P., Pérez-Quiñones, M.: Supporting student-written tests of GUI programs. ACM SIGCSE Bull. **40**(1), 537–541 (2008)
9. Ahmadzadeh, M., Janghorban, M., Jamasb, B.: JavaMarker extended: an Eclipse plugin to mark Java GUI programs. Int. J. Comput. Appl. **29**(10), 47–51 (2011)
10. Snyder, J., Edwards, S.H., Pérez-Quiñones, M.A.: LIFT: taking GUI unit testing to new heights. In: Proceedings of the 42nd ACM Technical Symposium on Computer Science Education, pp. 643–648 (2011)
11. Kruk, G., Alves, O., Molinari, L., Roux, E.: Best practices for efficient development of JavaFX applications. In: Proceedings of the 16th International Conference on Accelerator and Large Experimental Control Systems, pp. 1078–1083 (2017)
12. Klammer, C., Ramler, R., Stummer, H.: Harnessing automated test case generators for GUI testing in industry. In: Proceedings of the 42th Euromicro Conference on Software Engineering and Advanced Applications, pp. 227–234 (2016)
13. Akahane, Y., Kitaya, H., Inoue, U.: Design and evaluation of automated scoring Java programming assignments. In: Proceedings of the 16th IEEE/ACIS International Conference on Software Engineering, Artificial Intelligence, Networking and Parallel/Distributed Computing, pp. 1–6 (2015)

Python Deserialization Denial of Services Attacks and Their Mitigations

Kousei Tanaka and Taiichi Saito

Abstract In recent years, many vulnerabilities in deserialization mechanisms are reported. Serialization is converting an object to a byte string, and deserialization is converting the byte string to the object. Pickle is a serialization/deserialization module in Python standard library. In the pickle module, specially-crafted data consumes huge memory in deserialization. There is a possibility that the memory consumption leads to deniable of services attacks. This paper precisely describes the DoS attacks and their mitigations.

1 Introduction

Pickle is a serialization/deserialization module in Python standard library. In serialization (*pickling*), the pickle module converts a Python object into a byte string, and in deserialization (*unpickling*), it converts the byte string into the Python object.

This paper presents *denial of service* (DoS) attacks in deserialization in Python. While being unpickled, specially-crafted pickled data causes high memory consumption in the pickle module. This paper uses Python 3.5.2.

2 Related Researches

2.1 Java Deserialization Vulnerabilities

In recent years, many vulnerabilities for remote code execution in Java deserialization were reported. For example, Apache Commons Collections 3.2.1.4.0 have

K. Tanaka (✉) · T. Saito
Tokyo Denki University, Graduate School of Engineering, 5 Senju-Asahi-Cho,
Tokyo, Adachi-Ku 120-8551, Japan
e-mail: 17kmc05@ms.dendai.ac.jp

T. Saito
e-mail: taiichi@c.dendai.ac.jp

© Springer International Publishing AG, part of Springer Nature 2019
R. Lee (ed.), *Computational Science/Intelligence & Applied Informatics*, Studies
in Computational Intelligence 787, https://doi.org/10.1007/978-3-319-96806-3_2

15

remote code execution vulnerability in object deserialization [1]. On the other hand, some vulnerabilities for DoS attacks in Java deserialization were reported. Tomáš Polešovský presented a DoS attack, OIS-DoS [2, 3]. In OIS-DoS, specially-crafted serialized data is converted into a large array in deserialization, which consumes huge heap memory in Java Virtual Machine.

2.2 Python Deserialization Vulnerabilities

Marco Slaviero [4] showed a remote code execution vulnerability in unpickling. The pickle module does not verify pickled data before unpickling. When an object is deserialized in unpickling, any code in the constructor of the object is executed. On other hand, any DoS attacks in unpickling have not been reported to our best knowledge.

3 Background

This section describes the pickle module in Python standard library.

3.1 Pickle Module Warning

It has been known that Pickle library has remote code execution vulnerabilities. They are not fixed also in Python 3.6.4, and a document for Python 3.5.5 [5] warns as follows.

3.2 Pickle Protocols

The pickle module uses unique protocols called "pickle protocols", for pickling/unpickling. Table 1 shows the minimum Python version required each protocol. Higher-

Table 1 Pickle protocols

Pickle protocol	Overview	Required Python versions
0	Printable ASCII	Python 1 or Later
1	Binary format	Python 1 or Later
2	Efficient pickling	Python 2.3 or Later
3	Support bytes object	Python 3 or Later
4	Support large object	Python 3.4 or Later

type protocol requires newer Python version. Pickle protocol 0 is supported by all the versions of Python and generates pickled data that consists of printable ASCII characters. The latest version (Python 3.6.4) supports the five types of pickle protocols.

3.3 Pickle Virtual Machine's Overview

Pickled data is a program on a virtual pickle machine(PM, but more accurately called unpickling machine). It is a sequence of opcodes to be interpreted by PM and to build an arbitrarily complex Python object. It does not include looping, testing, or conditional instructions, or arithmetic or function calls. Opcodes are executed once each, from first to last, until a STOP opcode is reached. PM has two data areas, *the stack* and *the memo*.

Many opcodes push Python objects onto the stack; e.g., INT opcode pushes a Python integer object on the stack, whose value is gotten from a decimal string literal immediately following INT opcode in the pickle bytestream. Other opcodes take Python objects off the stack. The result of unpickling is whatever object is left on the top of the stack when STOP opcode is executed.

The memo is simply an array of objects, which is implemented as a dict object mapping little integers to objects. The memo serves as "long term memory in PM", and the little integers indexing the memo are akin to variable names. Some opcodes pop a stack object into the memo at a given index, and others push a memo object at a given index onto the stack [6].

3.4 Process of Unpickling

This section describes process of unpickling, by showing the result of unpickling pickled data "}." as an example. When the pickled data "}." is unpickled, load method, load_empty_dictionary method and load_stop method are used, which are included in the pickle module [7]. The following is the source code of load method. The load method repeats the procedure of reading an opcode from a pickled data file and calling the method corresponding to the opcode.

Source Code 1 load method

```
1  def load(self):
2    """Read a pickled object representation
3    from the open file.
4
5    Return the reconstituted object hierarchy
6    specified in the file.
7    """
8    # Check whether Unpickler was initialized
9    # correctly. This is
10   # only needed to mimic the behavior of
11   # _pickle.Unpickler.dump().
12   if not hasattr(self, "_file_read"):
13     raise UnpicklingError(
```

```
14      "Unpickler.__init__() was not called by "
15      "%s.__init__()" % (self.__class__.__name__,))
16  self._unframer = _Unframer(self._file_read,
17                             self._file_readline)
18  self.read = self._unframer.read
19  self.readline = self._unframer.readline
20  self.metastack = []
21  self.stack = []
22  self.append = self.stack.append
23  self.proto = 0
24  read = self.read
25  dispatch = self.dispatch
26  try:
27    while True:
28      key = read(1)
29      if not key:
30        raise EOFError
31      assert isinstance(key, bytes_types)
32      dispatch[key[0]](self)
33  except _Stop as stopinst:
34    return stopinst.value
```

The load method reads an opcode from the pickled data file and sets it to key. Then it calls the method corresponding to key. In this example, the load method first reads } opcode and calls the following load_empty_dictionary method.

Source Code 2 load_empty_dictionary method

```
1 def load_empty_dictionary(self):
2   self.append({})
3 dispatch[EMPTY_DICT[0]] = load_empty_dictionary
```

The load_empty_dictionary method pushes an empty dictionary onto the stack. Next, The load method reads '.' opcode and calls the following load_stop method.

Source Code 3 load_stop method

```
1 def load_stop(self):
2   value = self.stack.pop()
3   raise _Stop(value)
4 dispatch[STOP[0]] = load_stop
```

The load_stop method pops an object on the stack top, and the load method returns the object and finishes unpickling. In this example, the load method returns an empty dictionary object "{}".

4   DoS Attacks Using Unpickling

This section presents memory consumption DoS attacks that make PM deserialize specially-crafted pickled data. The pickled data forms a sequence of opcodes that push Python objects onto the stack in PM. During unpickling, since the pickled data continues to push Python objects onto the stack, the stack continues to grow deeper until STOP opcode, and finally consumes huge memory.

The opcode that pushes an empty dictionary onto the stack produces quite efficient attacks. The opcode that pushes an empty dictionary is one byte.

On other hand, the `empty dictionary` pushed onto the stack consumes 224 bytes memory. Pushed objects continue remaining on the stack until PM reads `STOP` opcode.

4.1 A DoS Attack

This subsection describes the details of a DoS attack using unpickling. This following code creates pickled data specially-crafted for DoS attack.

Source Code 4 creation-attack-code.py

```
1 #!/usr/bin/python3
2
3 num = 5 * 024 * 1024
4 pickle_data = b'}' * num
5 pickle_data += b'.'
6 with open('malicious.pickle','wb') as f:
7   f.write(pickle_data)
```

This code generates a pickled data file `malicious.pickle`. It consists of 5 * 1024 * 1024 repeats of the same opcode that pushes an `empty dictionary` object onto the stack and of one opcode `STOP`. When PM unpickles `malicious.pickle`, since an `empty dictionary` consumes 224 bytes memory in the stack, all generated `empty dictionary` objects consume about one gigabyte memory. These `empty dictionary` objects continue remaining in the stack until unpickling is finished.

4.2 Memory Consumption of Python Objects in the Stack

The script verification.py investigates memory consumption of Python objects in the stack in Python 3.5.2.

Source Code 5 verification.py

```
1 #!/usr/bin/python3
2
3 import tracemalloc
4 import logging
5 maxi = 5 * 1024 * 1024
6 cnt = 0
7 lis = []
8 logging.basicConfig(filename = 'tracemalloc.log',
9                      level = logging.DEBUG)
10 tracemalloc.start()
11 snapshot1 = tracemalloc.take_snapshot()
12 while cnt < num:
13   lis.append({})
14   cnt = cnt + 1
15   snapshot2 = tracemalloc.take_snapshot()
16   top_stats = snapshot2.compare_to(snapshot1,
17                        "lineno")
18   logging.debug("{0}{1}".format('='*50,
19                        len(lis)))
20   logging.debug("[Top 10 differences]")
21   for stat in top_stats[:10]:
22       logging.debug(stat)
```

This code reproduces the behavior of unpickling. Table 2, Table 3 and Table 4 show the results of running this code in Python 3.5.2 on Windows 10, Ubuntu 16.04 and CentOS 7, respectively.

These results say that the DoS attacks depend only on python environments and almost not on operating systems.

Next, the script estimate-deserializer.py investigates consumed memory of simple built-in objects in the stack by estimating memory consumption for 5 * 1024 * 1024 objects pushed onto the stack in Python 3.5.2.

Table 2 Windows 10

The number of pushed empty dictionaries	Memory size (B)	Memory block
10	2432	10
11	2656	11
12	2880	12
13	3104	13
14	3328	14
15	3552	15

Table 3 Ubuntu 16.04

The number of pushed empty dictionaries	Memory size (B)	Memory block
10	2368	10
11	2592	11
12	2816	12
13	3040	13
14	3264	14
15	3488	15

Table 4 CentOS 7

The number of pushed empty dictionaries	Memory size (B)	Memory block
10	2368	10
11	2592	11
12	2816	12
13	3040	13
14	3264	14
15	3488	15

Source Code 6 estimate-deserializer.py

```
1  #!/usr/bin/python3
2
3  import os
4  import sys
5  import subprocess
6  import shlex
7  import logging
8
9  logging.basicConfig(filename = 'output.log',
10                     level = logging.DEBUG)
11
12 def make_malicious_file():
13   NUM = 5 * 1024 * 1024
14   head = b'\x8f'
15   head = head * NUM
16   tail = b'.'
17   code = b''.join([head,tail])
18   with open('malicious.pickle','wb') as f:
19       f.write(code)
20
21 if __name__ == '__main__':
22   make_malicious_file()
23   p = subprocess.Popen('./deserializer.py',
24                        stdout=subprocess.PIPE)
25   command = "pmap -x " + str(p.pid) + " | tail -n 1"
26   while p.poll() == None:
27     monitor = subprocess.Popen(command,
28                               stdout=subprocess.PIPE,
29                               stderr=subprocess.PIPE,
30                               shell=True)
31     stdout_data,stderr_data = monitor.communicate()
32     logging.debug(stdout_data)
33   sys.exit(0)
```

Source Code 7 deserializer.py

```
1  #!/usr/bin/python3
2
3  import pickle
4
5  with open("malicious.pickle",mode="rb") as f:
6    pickle.load(f)
```

The script estimate-deserializer.py makes a pickled data file `malicious.pickle` and runs the script deserializer.py which unpickles `malicious.pickle`. It estimates memory consumption in deserializer.py. The result is shown in Table 5, which says that an `empty set` object has the largest amplification for memory consumption. An `empty set` object consumes 241 byte memory for one byte opcode '\x8f'.

However the opcode '\x8f' of pushing an `empty set` object onto the stack works only in pickle protocol 4 and is not included in ASCII. In an application that supports only pickle protocol with its version less than 4, or allows only human-readable ASCII characters, the opcode '\x8f' cannot be used for the DoS attack. On the other hand, since the opcode '}' of pushing an `empty dictionary` object onto the stack works in all versions of pickle protocols and is an ASCII character, the opcode '}' is available in wider environment than the opcode '\x8f'.

Table 5 Opcode object

Opcode name	Object type	Opcode string	Consuming memory (kb)	Pickle protocol
BININT	For-byte signed int	J\x00\x00\x00\x00	74,140	0
BININT1	1-byte unsigned int	K\x00	74,144	0
BINSTRING	Counted binary string	T\x01\x00\x00\x00a	74,144	0
SHORT_ BINSTRING	Counted binary string less 256 bytes	U\x01a	74,148	0
BINUNICODE	Unicode string; counted UTF-8 string	X\x01\x00\x00\x00a	32,176	0
EMPTY_DICT	empty_dictionary	}	1,197,300	0
EMPTY_LIST	empty_list	]	404,388	0
EMPTY_TUPLE	empty_tuple	)	74,148	0
TRUE	Boolean Object:TRUE	I01\n	74,148	0
FALSE	Boolean Object:FALSE	I00\n	74,148	0
NEWTRUE	Boolean Object:TRUE	\x88	74,148	2
NEWFALSE	Boolean Object:FALSE	\x89	74,144	2
EMPTY_SET	empty_set	\x8f	1,236,772	4

4.3 Impact

This subsection discusses the threat of the DoS attacks using pickle by considering the following attack scenario of a service that manages Python objects using the pickle module. At a scheduled maintenance time, the service pickles Python objects, stores the pickled data to a file and halts. At the scheduled restart time, it restarts and unpickles the pickled data from the file to restore the Python objects. If the file that includes the pickled data is replaced with a specially-crafted data file `malicious.pickle` during the maintenance, the service at the restart time unpickles `malicious.pickles`. It causes huge memory consumption and may fail to restart.

Warning: The pickle module is not secure against erroneous or maliciously constructed data. Never unpickle data received from an untrusted or unauthenticated source.

Fig. 1 Pickle module warning

Table 6 Pickle and pickletools results

Consumed memory of unpickling (kb)	Consumed memory of disassembling (kb)
1,236,060	154,348

4.4 Mitigations

This subsection discusses mitigations for the DoS attacks. The most simple mitigation is not to unpickle data received from an untrusted or unauthenticated source as described Fig. 1.

Pickletools module defines some functions for analyzing pickled data. The `pickletools.dis` function disassembles and analyzes pickled data and does not unpickle it, so it consumes less memory than actual unpickling.

When `pickletools.dis` function reads STOP opcode, if there remain two or more objects in the stack, `pickletools.dis` function determines the pickled data invalid and outputs error code. For example, if `pickletools.dis` function reads `malicious.pickle`, it outputs "ValueError: stack not empty after STOP: [dict]". The DoS attacks can be mitigated by using the `pickletools.dis` function before unpickling. Table 6 compares consumed memory of unpickling `malicious.pickle` and that of disassembling `malicious.pickle` using pickletools.

5 Conclusion

This paper reported memory consumption DoS attacks in unpickling in the pickle module. In the DoS attack, PM pushes a lot of `empty dictionary` objects onto the stack and continues remaining the objects in the stack until STOP opcode. An `empty set` object consumes 241 byte memory in the stack. An opcode of pushing an `empty set` object onto the stack is one byte. So, the DoS attacks amplification is 241. The DoS attacks can be mitigated by using the `pickletools.dis` function before unpickling.

Appendix

Reply from Python Team

This subsection describes Python team's reply to our report on the DoS attacks (Fig. 2).

Changelog of Python Built-in Objects

This subsection describes the change history of Python built-in objects. Our DoS attacks were found in Python 3.5.2 and reported to Python team. However the changelog of the version 3.6.0 says as follows (Fig. 3).

This changelog is in python 3.6.0. An `empty dictionary` has reduced consumption memory with this change. However, an `empty set` consumes huge memory as in Python 3.5.2. The DoS attacks are still possible.

Thanks for your report.

I do not believe this is a vulnerability – a program that reads pickles from untrusted sources has bigger problems (it's easy to send a pickle that runs arbitrary code).

Fig. 2 Python team's reply

New dict implementation

The dict type now uses a "compact" representation based on a proposal by Raymond Hettinger which was first implemented by PyPy. The memory usage of the new dict() is between 20% and 25% smaller compared to Python 3.5.

The order-preserving aspect of this new implementation is considered an implementation detail and should not be relied upon (this may change in the future, but it is desired to have this new dict implementation in the language for a few releases before changing the language spec to mandate order-preserving semantics for all current and future Python implementations; this also helps preserve backwards-compatibility with older versions of the language where random iteration order is still in effect, e.g. Python 3.5).

(Contributed by INADA Naoki in bpo-27350. Idea originally suggested by Raymond Hettinger.)

Fig. 3 Dict object

References

1. Collections—Commons Collections Security Reports. https://commons.apache.org/proper/commons-collections/security-reports.html. Accessed 31 March 2018
2. Tomáš Polešovský. http://topolik-at-work.blogspot.jp/2016/04/java-deserialization-dos-payloads.html. Accessed 23 March 2018
3. Java-Deserialization-Cheat-Sheet. https://github.com/topolik/ois-dos/. Accessed 23 March 2018
4. Marco Slaviero—Sour Pickle. https://media.blackhat.com/bh-us-11/Slaviero/BH_US_11_Slaviero_Sour_Pickles_Slides.pdf. Accessed 23 March 2018
5. 12.1. Pickle Python object serialization—Python 3.5.5 documentation. https://docs.python.org/3.5/library/pickle.html#module-pickle. Accessed 23 March 2018
6. cpython/pickletools.py at master—python/cpython. https://github.com/python/cpython/blob/master/Lib/pickletools.py. Accessed 23 March 2018
7. cpython/pickle.py at master—python/cpython. https://github.com/python/cpython/blob/master/Lib/pickle.py. Accessed 23 March 2018

A Branch-and-Bound Based Exact Algorithm for the Maximum Edge-Weight Clique Problem

Satoshi Shimizu, Kazuaki Yamaguchi and Sumio Masuda

Abstract The maximum edge-weight clique problem is to find a clique whose sum of edge-weight is maximum for a given edge-weighted undirected graph. The problem is NP-hard and was formulated as a mathematical programming problem in previous studies. In this paper, we propose an exact algorithm based on branch-and-bound. By some computational experiments, we confirmed our proposal algorithm is faster than the methods based on mathematical programming.

Keywords Maximum edge-weight clique problem · NP-hard
Branch-and-bound · Upper bound calculation

1 Introduction

A clique in an undirected graph is a set of vertices C such that any two vertices in C are adjacent. The maximum clique problem (MCP) is to find a clique of maximum cardinality in a given undirected graph. Given an undirected graph $G = (V, E)$ and non-negative weight $w(v)$ to each vertex $v \in V$, the maximum weight clique problem (MWCP) is to find a clique C of maximum sum of vertex weights. Given an undirected graph $G = (V, E)$ and non-negative weight $w_e(u, v)$ to each edge $(u, v) \in E$, the maximum edge-weight clique problem (MEWCP) is to find a clique of maximum sum of edge weights [21]. Given an edge-weighted complete graph and a natural number b, the maximum diversity problem (MDP) [17], also known as b-clique problem [25], is to find a clique of size b that has maximum sum of edge weights. Note that MDP is different from MEWCP of our definition although MDP may sometimes be

S. Shimizu · K. Yamaguchi (✉) · S. Masuda
Kobe University Faculty of Engineering, 1-1 Rokkodai, Nada, Kobe 657-8501, Japan
e-mail: ky@kobe-u.ac.jp

S. Shimizu
e-mail: ss81054@gmail.con

S. Masuda
e-mail: masuda@kobe-u.ac.jp

© Springer International Publishing AG, part of Springer Nature 2019
R. Lee (ed.), *Computational Science/Intelligence & Applied Informatics*, Studies
in Computational Intelligence 787, https://doi.org/10.1007/978-3-319-96806-3_3

called MEWCP [1]. MCP and its generalizations, MWCP and MEWCP, are known to be NP-hard [12]. There are lots of applications of MCP, MWCP and MEWCP: coding theory [5], network design [26], computer vision [14], bioinformatics [15], auctions [7], etc. By introducing edge-weights, more information can be obtained for the following applications: protein side-chain packing [3, 6], market basket analysis [8], communication analysis [9, 10] and so on.

For MCP and MWCP, a number of exact algorithms based on branch-and-bound have been proposed. Branch-and-bound consists of a branching procedure and a bounding procedure. A branching procedure divides a problem into smaller sub-problems and solves them in a recursive manner. A bounding procedure calculates an upper bound for each subproblem, and then prunes the subproblem if it does not contain global optimum solutions. We briefly introduce some bounding procedures below. For MCP, algorithms in [27, 28] and their variants [4, 22] use bounding procedures based on heuristic vertex coloring. Recently dynamic vertex ordering to reduce search space is proposed [16]. For MWCP, a bounding procedure in [19] runs in $O(1)$ time using optimum values of subproblems already searched. In [30], longest paths in a directed acyclic graph are calculated in a bounding procedure. A bounding procedure in [24] uses optimal tables constructed before the branch-and-bound. An upper bound calculation based on MaxSAT reasoning is proposed in [11]. Some of these algorithms are compared in [18].

For MDP, some formulations in mathematical programming [2, 20, 25] and a branch-and-bound based algorithm [17] have been proposed. On the other hand, for MEWCP, there are only exact methods [13, 23] that formulate MEWCP in mathematical programming.

In this paper, we propose an exact algorithm for MEWCP based on branch-and-bound. The bounding procedure of our algorithm decomposes subproblems into three components and calculates an upper bound of each component. By some experiments, we confirmed that our algorithm can obtain exact solutions in shorter time than the methods based on mathematical programming.

The remainder of this paper is organized as follows. Formulations in mathematical programming are shown in Sect. 2. Some branch-and-bound algorithms for MWCP are described in Sect. 3. Our algorithm *EWCLIQUE* is described in Sect. 4. The results of computational experiments are in Sect. 5. We conclude the paper in Sect. 6.

2   Formulations in Mathematical Programming

In this section, we introduce mathematical programming formulations of MEWCP: Quadratic Programming (QP), Integer Programming (IP) and Mixed Integer Programming (MIP) [23]. Let $V = \{v_1, v_2, \ldots, v_n\}$ and $\bar{E} = \{(v_i, v_j) \mid i \neq j, (v_i, v_j) \notin E\}$. By the condition $i \neq j$, $\bar{E}$ does not include self loops.

2.1 Quadratic Programming

MEWCP can be formulated in QP as follows:

$$\text{maximize:} \sum_{(v_i, v_j) \in E} w_e(v_i, v_j) x_i x_j \tag{1}$$

$$\text{s.t.:}\ x_i + x_j \le 1, \forall (v_i, v_j) \in \bar{E} \tag{2}$$

$$x_i \in \{0, 1\}, \forall v_i \in V. \tag{3}$$

A binary variable x_i is set to 1 if and only if v_i is in a solution. By constraint (2) at most one of any nonadjacent pair of vertices can be in a clique.

2.2 Integer Programming

MEWCP can be formulated in IP as follows:

$$\text{maximize:} \sum_{(v_i, v_j) \in E} w_e(v_i, v_j) y_{ij} \tag{4}$$

$$\text{s.t.:}\ y_{ij} \le x_i, \forall (v_i, v_j) \in E \tag{5}$$

$$y_{ij} \le x_j, \forall (v_i, v_j) \in E \tag{6}$$

$$x_i + x_j \le 1, \forall (v_i, v_j) \in \bar{E} \tag{7}$$

$$x_i \in \{0, 1\}, \forall v_i \in V \tag{8}$$

$$y_{ij} \in \{0, 1\}, \forall (v_i, v_j) \in E. \tag{9}$$

A binary variable x_i is set to 1 if and only if v_i is in a solution. By constraints (5) and (6), for any edge (v_i, v_j), a binary variable y_{ij} is set to 1 only when both v_i and v_j are in the solution.

2.3 Mixed Integer Programming

$N(v)$ denotes the set of all vertices adjacent to v. Let $N^+(v_i) = N(v_i) \cap \{v_j \mid j > i\}$. Let $U_i = \sum_{v_j \in N^+(v_i)} w_e(v_i, v_j)$. MEWCP can be formulated in MIP as follows:

$$\text{maximize:} \sum_{v_i \in V \setminus \{v_n\}} z_i \tag{10}$$

$$\text{s.t.:} \; x_i + x_j \leq 1, \forall (v_i, v_j) \in \bar{E} \tag{11}$$

$$z_i \leq U_i x_i, \forall v_i \in V \setminus \{v_n\} \tag{12}$$

$$z_i \leq \sum_{v_j \in N^+(v_i)} w_e(v_i, v_j) x_j, \forall v_i \in V \setminus \{v_n\} \tag{13}$$

$$x_i \in \{0, 1\}, \forall v_i \in V. \tag{14}$$

A binary variable x_i is set to 1 if and only if v_i is in a solution. If $x_i = 0$, $z_i = 0$ by the constraint (12). If $x_i = 1$, the value of z_i is determined by the constraints (13) because the constraint (12) is looser than the constraint (13) in this case. By the constraint (13), z_i is bounded by the edge-weight sum of edges in the solution whose endpoints are v_i and $v_j \in N^+(v_i)$.

3 Branch-and-Bound for MWCP

In this section, two known branch-and-bound based algorithms for MWCP are described. Our algorithm for MEWCP is based on these algorithms. For $V' \subseteq V$, let $W(V') = \sum_{v \in V'} w(v)$. $P_{MWC}(C, S)$ denotes a subproblem of MWCP, where C is a set of vertices already chosen and S is a set of candidate vertices to be added to C. Note that $C \cap S$ must be an empty set, and for any element $v \in S, C \subseteq N(v)$ must be satisfied. The problem corresponding to the given graph $G = (V, E)$ is $P_{MWC}(\emptyset, V)$.

In a branching procedure, a subproblem $P_{MWC}(C, S)$ is divided into $|S|$ sub-problems and examined in depth-first recursive manner. For each subproblem, a bounding procedure calculates an upper bound for the weight of feasible solutions in $P_{MWC}(C, S)$. By using the upper bounds, a bounding procedure prunes unnecessary subproblems to reduce the search tree size. The tightness and calculation time for each upper bound are important for the entire computation time of branch-and-bound.

3.1 Östergård's Algorithm

A branch-and-bound algorithm for MWCP is proposed by Östergård [19]. Let $V_i = \{v_i, v_{i-1}, \ldots, v_1\}$. Östergård's algorithm calculates exact solutions of $P_{MWC}(\emptyset, V_1)$, $P_{MWC}(\emptyset, V_2), \ldots, P_{MWC}(\emptyset, V_n)$ in this order. An exact solution of given instance is finally obtained because $P_{MWC}(\emptyset, V_n)$ is equivalent to the given problem. Each time exact solution of $P_{MWC}(\emptyset, V_i)$ is obtained, its weight is stored in $c[i]$. The array $c[\cdot]$ is used to calculate upper bounds in the following way. Hereafter $M(V')$ denotes the maximum index of vertices in V'. Let F be a feasible solution of $P_{MWC}(C, S)$. Since $S \subseteq V_{M(S)}$, the following inequality holds :

$$W(F) = W(C) + W(S \cap F) \qquad (15)$$
$$\leq W(C) + W(V_{M(S)} \cap F) \qquad (16)$$
$$\leq W(C) + c[M(S)]. \qquad (17)$$

3.2 Longest Path Method

An upper bound calculation for MWCP is proposed in [30]. Let $G(V')$ be the sub-graph of G induced by $V' \subseteq V$. For a vertex induced subgraph $G(S)$, let $\mathbf{D}$ be a directed acyclic graph such that the graph obtained by replacing each directed edge in $\mathbf{D}$ with an undirected edge is isomorphic to $G(S)$. The *length of a path* is defined as the sum of vertex weights in the path. Let $L(\mathbf{D})$ be the length of a longest path in $\mathbf{D}$. Let F be a feasible solution of $P_{MWC}(C, S)$. Since $F \cap S$ is a clique in $G(S)$, at least one path in $\mathbf{D}$ includes all vertices of $F \cap S$ and the following inequality holds:

$$W(F) = W(C) + W(S \cap F) \qquad (18)$$
$$\leq W(C) + L(\mathbf{D}). \qquad (19)$$

Hence an upper bound for $W(F)$ can be obtained from a longest path in a directed acyclic graph $\mathbf{D}$. It is called *longest path method*.

4 Our Algorithm *EWCLIQUE*

4.1 Branch-and-Bound for MEWCP

Our algorithm EWCLIQUE is the first algorithm for MEWCP based on branch-and-bound. Before introducing EWCLIQUE, we show the representation of subproblems of MEWCP. Hereafter, for a clique $C \subseteq V$, let $W_e(C) = \sum_{u,v \in C} w_e(u, v)$.

4.1.1 Subproblems of MEWCP

$P_{MEWC}(C, S)$ denotes a subproblem of the MEWCP, where C is a set of vertices already chosen and S is a set of candidate vertices to be added to C. Note that $C \cap S$ must be an empty set, and for any element $v \in S$, $C \subseteq N(v)$ must be satisfied. The problem corresponding to the given graph $G = (V, E)$ is $P_{MEWC}(\emptyset, V)$.

Figure 1a shows an example of edge-weighted graph G_{ex}, and Fig. 1b shows the edge-weight matrix of G_{ex}. In the edge-weight matrix, the blank spaces denote that vertices are not adjacent. Figure 2 shows a subproblem $P_{MEWC}(C_{ex}, S_{ex})$ of the graph G_{ex}.

Fig. 1 A graph example G_{ex}

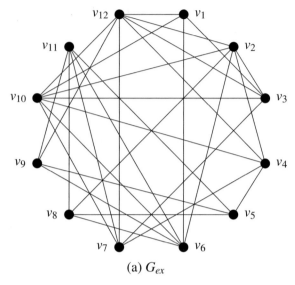

(a) G_{ex}

	v_1	v_2	v_3	v_4	v_5	v_6	v_7	v_8	v_9	v_{10}	v_{11}	v_{12}
v_1	-	3			2					1		4
v_2		-	2	1		1		2		2		2
v_3	3	2	-				4			5		3
v_4		1		-	4		6			3		7
v_5				4	-			4	6		1	
v_6	2	1				-		4	3	5	6	
v_7			4	6			-			3	4	6
v_8		2			4	4		-			5	
v_9					6	3			-		10	5
v_{10}	1	2	5	3		5	3			-		6
v_{11}					1	6	4	5	10		-	
v_{12}	4	2	3	7			6		5	6		-

(b) Edge-weight matrix of G_{ex}

4.1.2 Three Components of a Feasible Solution

Let F be a feasible solution of $P_{MEWC}(C, S)$. Figure 3 shows the relationship between F and $P_{MEWC}(C, S)$. Since $F = C \cup (S \cap F)$, edges in $G(F)$ can be decomposed into the following three groups:

- Edges between two vertices in C.
- Edges between a vertex in C and a vertex in $S \cap F$.
- Edges between two vertices in $S \cap F$.

According to the decomposition, the following equation is obtained:

Fig. 2 A subproblem $P_{MEWC}(C_{ex}, S_{ex})$ of G_{ex}

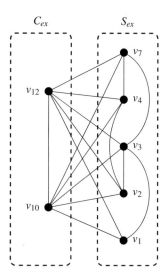

Fig. 3 F in $P_{MEWC}(C, S)$

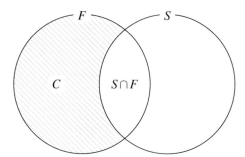

$$W_e(F) = W_e(C) + \sum_{u \in C} \sum_{v \in S \cap F} w_e(u, v) + W_e(S \cap F). \qquad (20)$$

Figure 4 shows the three components of $W(F)$. For the second term of the Eq. (20), no corresponding terms are in the equations for MWCP (15)–(19). To adapt branch-and-bound to MEWCP, efficient upper bound calculation methods for these terms are required.

4.2 Outline of EWCLIQUE

The outline of our algorithm EWCLIQUE is as follows:

Branching Procedure
    Similar to Östergård's algorithm for MWCP, EWCLIQUE finds optimum solutions of $P_{MEWC}(\emptyset, V_1)$, $P_{MEWC}(\emptyset, V_2)$, ..., $P_{MEWC}(\emptyset, V_n)$ in this order, where

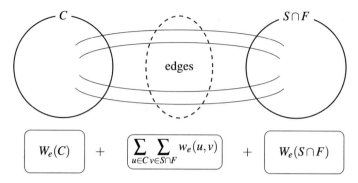

Fig. 4 Components of $W(F)$ calculation

V_i denotes a vertex set $\{v_i, v_{i-1}, \ldots, v_1\}$. It solves each $P_{MEWC}(\emptyset, V_i)$ by branch-and-bound and stores each optimum value of $P_{MEWC}(\emptyset, V_i)$ in $c[i]$ to use in the bounding procedure.

Bounding Procedure

EWCLIQUE decomposes a subproblem into three components described in Sect. 4.1.2, and calculates an upper bound of each component. For the third term $W_e(S \cap F)$ of the Eq. (20), the array $c[\cdot]$ calculated in the branching procedure is used. To handle the second term $\sum_{u \in C} \sum_{v \in S \cap F} w_e(u, v)$ of the Eq. (20), EWCLIQUE introduces a pseudo vertex weights described in Sect. 4.2.1.

4.2.1 Pseudo Vertex Weights

For each vertex $v \in S$ of a subproblem $P_{MEWC}(C, S)$, EWCLIQUE introduces a pseudo vertex weight $w_\rho(C, v) = \sum_{u \in C} w_e(u, v)$. For example, Fig. 5 shows the pseudo vertex weight of each $v \in S_{ex}$. Hereafter, $w_\rho(v)$ denotes $w_\rho(C, v)$ when $P_{MEWC}(C, S)$ can be obviously identified.

The pseudo vertex weights satisfy the following equation:

$$\sum_{u \in C} \sum_{v \in S \cap F} w_e(u, v) = \sum_{v \in S \cap F} w_\rho(v). \tag{21}$$

Hence, after assigning $w_\rho(v)$ to each $v \in S$, EWCLIQUE can calculate an upper bound of the second term $\sum_{u \in C} \sum_{v \in S \cap F} w_e(u, v)$ of the Eq. (20) only with $G(S \cap F)$, without C. Further details of the upper bound calculation is shown in Sect. 4.5.

Fig. 5 Pseudo vertex weights for P_{MEWC} (C_{ex}, S_{ex})

$v_i \in S_{ex}$	$w_\rho(v_i)$
v_1	$w_e(v_1, v_{10}) + w_e(v_1, v_{12}) = 5$
v_2	$w_e(v_2, v_{10}) + w_e(v_2, v_{12}) = 4$
v_3	$w_e(v_3, v_{10}) + w_e(v_3, v_{12}) = 8$
v_4	$w_e(v_4, v_{10}) + w_e(v_4, v_{12}) = 10$
v_7	$w_e(v_7, v_{10}) + w_e(v_7, v_{12}) = 9$

4.3 Main Routine

We show in Algorithm 1 the main routine of our algorithm EWCLIQUE. Before branch-and-bound, EWCLIQUE renumbers vertex indexes for efficiency (described in Sect. 4.6). The global variable C_{max} is the current-best solution, initialized by an empty set. In the **for** loop, EWCLIQUE finds optimum solutions of $P_{MEWC}(\emptyset, V_1)$, $P_{MEWC}(\emptyset, V_2)$, ..., $P_{MEWC}(\emptyset, V_n)$ by the subroutine EXPAND described in Sect. 4.4. In each loop, the subroutine EXPAND updates C_{max} if better solution is found, and the value of $W_e(C_{max})$ is stored in $c[i]$ for calculation of upper bounds of subproblems.

Algorithm 1 EWCLIQUE

INPUT: $G = (V, E), w_e(\cdot, \cdot)$
OUTPUT: a maximum edge-weight clique C_{max}
GLOBAL VARIABLES: $C_{max}, c[\cdot]$
1: Renumber vertex indexes. (described in Sect. 4.6)
2: $C_{max} \leftarrow \emptyset$
3: **for** i from 1 to $|V|$ **do**
4:     EXPAND($\emptyset, V_i$)   ▷ C_{max} is updated if better solution is found (described in Sect. 4.4)
5:     $c[i] \leftarrow W_e(C_{max})$
6: **end for**
7: **return** C_{max}

Algorithm 2 Solving a subproblem

INPUT: a subproblem $P_{MEWC}(C, S)$
OUTPUT: Update C_{max} to a better clique if it exists.
GLOBAL VARIABLES: $C_{max}, c[\cdot]$
1: **procedure** EXPAND(C, S)
2:     **if** $S = \emptyset$ **then**
3:         **if** $W_e(C) > W_e(C_{max})$ **then**
4:             $C_{max} \leftarrow C$
5:         **end if**
6:         **return**
7:     **end if**
8:     $lp[\cdot] \leftarrow$ LONGESTPATH(C, S)   ▷ described in Sect. 4.5.2
9:     **while** $S \neq \emptyset$ **do**
10:         **if** $W_e(C) + c[M(S)] + lp[M(S)] > W_e(C_{max})$ **then**
11:             $C' \leftarrow C \cup \{v_{M(S)}\}$
12:             $S' \leftarrow S \cap N(v_{M(S)})$
13:             EXPAND(C', S')   ▷ $P_{MEWC}(C', S')$
14:         **end if**
15:         $S \leftarrow S \setminus \{v_{M(S)}\}$
16:     **end while**
17: **end procedure**

4.4 Subroutine EXPAND

Here we describe the detail of the subroutine EXPAND (in Algorithm 2), which calls itself recursively to find better solution than current C_{max}. In the case of $S = \emptyset$, EXPAND updates C_{max} if $W_e(C) > W_e(C_{max})$ (lines 3–5), and it is the base case of recursive calls (line 9). If $S \neq \emptyset$, in the while loop from line 9, EXPAND divides $P_{MEWC}(C, S)$ into $|S|$ subproblems and examines them recursively.

For each loop from line 9, a subproblem $P_{MEWC}(C', S')$ is created. C' is a clique obtained by adding $v_{M(S)}$ to C (line 11). S' is a candidate vertex set obtained by removing non-adjacent vertices of $v_{M(S)}$ from S (line 12). At the end of the while loop, $v_{M(S)}$ is removed from S and continue the loop unless $S = \emptyset$.

During this process, EWCLIQUE calculates upper bounds of feasible solutions and prunes unnecessary subproblems in line 10. The subroutine "LONGESTPATH" in line 8 constructs a directed acyclic graph from $P_{MEWC}(C, S)$, calculates longest paths of the DAG using pseudo vertex weights, and store their length in the array $lp[\cdot]$ of size $M(S)$. The array $lp[\cdot]$ is used for upper bound calculation.

4.5 Upper Bound Calculation

Let F' be a feasible solution of $P_{MEWC}(C', S')$ in line 13 of Algorithm 2. Here we show that the value of $W_e(C) + c[M(S)] + lp[M(S)]$ in line 10 is an upper bound of $W_e(F')$. From the lines 11 and 12 of Algorithm 2, $C' = C \cup \{v_{M(S)}\}$ and $S' = S \cap N(v_{M(S)})$. Figure 6 schematically illustrates inclusion relation of $P_{MEWC}(C, S)$, $P_{MEWC}(C', S')$ and F' by intervals.

Since $F' = C \cup (S \cap F')$, the following equation holds similarly to (20):

$$W_e(F') = W_e(C) + \sum_{u \in C} \sum_{v \in S \cap F'} w_e(u, v) + W_e(S \cap F'). \tag{22}$$

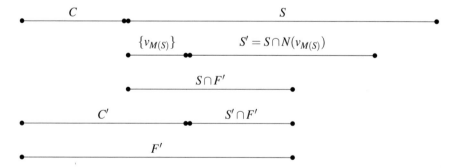

Fig. 6 Inclusion relation of $P_{MEWC}(C, S)$, $P_{MEWC}(C', S')$ and F'

EWCLIQUE calculates an upper bound of the Eq. (22). The accurate value of $W_e(C)$ is already calculated during the branching procedure. Therefore our algorithm calculates the following upper bounds:

$UB1$: an upper bound for $W_e(S \cap F')$
$UB2$: an upper bound for $\sum_{u \in C} \sum_{v \in S \cap F'} w_e(u, v)$.

$W_e(C) + UB1 + UB2$ is an upper bound for $W_e(F')$ and used to prune unnecessary subproblems. The calculation of $UB1$ is described in Sect. 4.5.1. The calculation of $UB2$ is described in Sect. 4.5.2.

4.5.1 UB1: Upper Bound of $W(S \cap F')$

Since $S \subseteq V_{M(S)}$, following inequality holds:

$$W_e(S \cap F') \leq W_e(V_{M(S)} \cap F') \leq c[M(S)]. \tag{23}$$

Therefore $c[M(S)]$ can be used as $UB1$. For the tightness of upper bounds, in the subroutine EXPAND, the largest indexed vertex in C is chosen as a branching variable, so that vertices of large indexes disappear from S in early stages (line 13, Algorithm 2).

4.5.2 UB2: Upper Bound of $\sum_{u \in C} \sum_{v \in S \cap F'} w_e(u, v)$

For a subproblem $P_{MEWC}(C, S)$, EWCLIQUE constructs a vertex-weighted directed acyclic graph, and calculates the length of longest paths that can be used as $UB2$. $(\overrightarrow{v, u})$ denotes a directed edge from v to u when it is necessary to distinguish from an undirected edge, and the vertex-weighted directed acyclic graph consists of follows:

- A vertex set S.
- A directed edge set $\{(\overrightarrow{v_i, v_j}) \mid (v_i, v_j) \in E(S), i < j\}$, where $E(S)$ is the edge set of $G(S)$.
- A pseudo vertex weight $w_\rho(\cdot)$ calculated from $P_{MEWC}(C, S)$ is assigned as a vertex weight to each vertex in S.

$\mathbf{D}(C, S)$ denotes this vertex-weighted directed acyclic graph for $P_{MEWC}(C, S)$. Hereafter $\mathbf{D}$ denotes $\mathbf{D}(C, S)$, when $P_{MEWC}(C, S)$ can be obviously identified. Figure 7 shows $\mathbf{D}_{ex}$ of $P_{MEWC}(C_{ex}, S_{ex})$. Numbers in parentheses represent vertex weights.

Let $[v_1, \ldots, v_k]$ a path from v_1 to v_k. We define path length in $\mathbf{D}$ as the sum of vertex weights in the path. Let $\Pi(\mathbf{D}, v)$ be a set of paths in $\mathbf{D}$ whose endpoint is v. Let $L(\mathbf{D}, v)$ be length of longest paths in $\Pi(\mathbf{D}, v)$. For example, $\Pi(\mathbf{D}_{ex}, v_3) = \{[v_1, v_3], [v_2, v_3]\}$. The longest path in $\Pi(\mathbf{D}_{ex}, v_3)$ is $[v_1, v_3]$ and its length is $L(\mathbf{D}_{ex}, v_3) = 13$, respectively. Note that any path in $\Pi(\mathbf{D}, v_i)$ does not include vertices in $V \setminus V_i$ because of the edge direction of $\mathbf{D}$. Since F' is a feasible solution of $P_{MEWC}(C', S')$ where $C' = C \cup \{v_{M(S)}\}$ and $S' = S \cap N(v_{M(S)})$, F' is a clique including $v_{M(S)}$. Hence there

Fig. 7 $\mathbf{D}_{ex}$

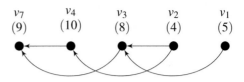

Algorithm 3 Calculate length of longest paths

INPUT: a subproblem $P_{MEWC}(C, S)$
OUTPUT: an array $lp[\cdot]$ that contains length of longest paths
1: **procedure** LONGESTPATH(C, S)
2:     $S' \leftarrow S$
3:     **while** $S' \neq \emptyset$ **do**
4:         $i \leftarrow \min\{j \mid v_j \in S'\}$
5:         **if** $S \cap V_i \cap N(v_i) = \emptyset$ **then**
6:             $lp[i] \leftarrow w_\rho(v_i)$
7:         **else**
8:             $lp[i] \leftarrow w_\rho(v_i) + \max\{lp[u] \mid u \in S \cap V_i \cap N(v_i)\}$
9:         **end if**
10:        $S' \leftarrow S' \setminus \{v_i\}$
11:    **end while**
12:    **return** $lp[\cdot]$
13: **end procedure**

---

exists at least one path in $\Pi(\mathbf{D}, v_{M(S)})$ that includes all vertices of $S \cap F'$. Therefore following inequality holds:

$$\sum_{u \in C} \sum_{v \in S \cap F'} w_e(u, v) = \sum_{v \in S \cap F'} w_\rho(v) \leq L(\mathbf{D}, v_{M(S)}). \tag{24}$$

Our algorithm uses $L(\mathbf{D}, v_{M(S)})$ as $UB2$. At line 8 of Algorithm 2, the subroutine LONGESTPATH shown in Algorithm 3 is called to calculate $UB2$. For each $v_i \in S$, it calculates $L(\mathbf{D}, v_i)$ and stores the value with $lp[i]$. The calculation of $lp[i]$ is done in nonincreasing order of i. There are two cases of calculating $lp[i]$. One is the case that no neighbor of v_i is in $S \cap V_i$ (line 6). In this case, $\Pi(\mathbf{D}, v_i)$ is $\{[v_i]\}$ because of the edge direction of $\mathbf{D}$. Hence $lp[i]$ equals to $w_\rho(v_i)$. In the other case that some neighbors of v_i are in $S \cap V_i$ (line 8), $lp[i]$ equals to the sum of $w_\rho(v_i)$ and the length of longest paths connected to v_i.

4.5.3   A Case Study

Here is an example of upper bound calculation for $P_{MEWC}(C_{ex}, S_{ex})$ of Fig. 2. Figure 8 shows $c[\cdot]$ that are already calculated. First our algorithm calculates $lp[i]$ for each $v_i \in S_{ex}$ using $w_\rho(v_i)$ and $\mathbf{D}_{ex}$ in Figs. 5 and 7 respectively. For v_1, $lp[v_1] = w_\rho(v_1)$ because $S_{ex} \cap V_1 \cap N(v_1) = \emptyset$. Same as $lp[v_1]$, $lp[v_2] = w_\rho(v_2)$. For v_3, $S_{ex} \cap V_3 \cap N(v_3) = \{v_1, v_2\} \neq \emptyset$. Our algorithm selects v_1 from $\{v_1, v_2\}$

Fig. 8 Optimal weights stored in $c[\cdot]$

subproblem	optimal weight
$P_{MEWC}(\emptyset, V_1)$	$c[1] = 0$
$P_{MEWC}(\emptyset, V_2)$	$c[2] = 0$
$P_{MEWC}(\emptyset, V_3)$	$c[3] = W_e(\{v_1, v_3\}) = 3$
$P_{MEWC}(\emptyset, V_4)$	$c[4] = W_e(\{v_1, v_3\}) = 3$
$P_{MEWC}(\emptyset, V_5)$	$c[5] = W_e(\{v_4, v_5\}) = 4$
$P_{MEWC}(\emptyset, V_6)$	$c[6] = W_e(\{v_4, v_5\}) = 4$
$P_{MEWC}(\emptyset, V_7)$	$c[7] = W_e(\{v_4, v_7\}) = 6$
$P_{MEWC}(\emptyset, V_8)$	$c[8] = W_e(\{v_2, v_6, v_8\}) = 7$
$P_{MEWC}(\emptyset, V_9)$	$c[9] = W_e(\{v_2, v_6, v_8\}) = 7$
$P_{MEWC}(\emptyset, V_{10})$	$c[10] = W_e(\{v_4, v_7, v_{10}\}) = 12$
$P_{MEWC}(\emptyset, V_{11})$	$c[11] = W_e(\{v_6, v_9, v_{11}\}) = 19$
$\vdots$	$\vdots$

Fig. 9 Longest path calculation

v_i	$S_{ex} \cap V_i \cap N(v_i)$	$lp[i]$	
v_1	$\emptyset$	$w_\rho(v_1)$	$= 5$
v_2	$\emptyset$	$w_\rho(v_2)$	$= 4$
v_3	$\{v_1, v_2\}$	$w_\rho(v_3) + \max(lp[v_1], lp[v_2])$	$= 13$
v_4	$\{v_2\}$	$w_\rho(v_4) + \max(lp[v_2])$	$= 14$
v_7	$\{v_3, v_4\}$	$w_\rho(v_7) + \max(lp[v_3], lp[v_4])$	$= 23$

that constructs the longest path. Therefore $lp[v_3] = w_\rho(v_3) + \max(lp[v_1], lp[v_2]) = w_\rho(v_3) + lp[v_1] = 13$. Similarly, for v_4 and v_7, length of paths are calculated. The result is shown in Fig. 9.

Let F'_{ex} be a feasible solution of the subproblem $P_{MEWC}(C'_{ex}, S'_{ex})$ such that $S'_{ex} = S_{ex} \cap N(v_7) = \{v_3, v_4\}$ and $C'_{ex} = C_{ex} \cup \{v_7\} = \{v_{12}, v_{10}, v_7\}$. Since $S_{ex} \subseteq V_{M(S_{ex})} = V_7$, $UB1$ can be calculated as following:

$$W_e(S_{ex} \cap F'_{ex}) \le W_e(V_7 \cap F'_{ex}) \tag{25}$$

$$\le c[7] \tag{26}$$

$$= 6. \tag{27}$$

Since $M(S_{ex}) = 7$, $UB2$ is calculated as follows:

$$\sum_{u \in C_{ex}} \sum_{v \in S_{ex} \cap F'_{ex}} w_e(u, v) \le L_\rho(\mathbf{D}_{ex}, v_7) \tag{28}$$

$$= lp[v_7] \tag{29}$$

$$= 23. \tag{30}$$

In summary, an upper bound for $W_e(F'_{ex})$ is calculated as follows:

$$W_e(F'_{ex}) = W_e(C_{ex}) + \sum_{u \in C_{ex}} \sum_{v \in S_{ex} \cap F'_{ex}} w_e(u, v) + W_e(S_{ex} \cap F'_{ex}) \qquad (31)$$

$$\leq W_e(C_{ex}) + lp[7] + c[7] \qquad (32)$$

$$= 6 + 23 + 6 \qquad (33)$$

$$= 35. \qquad (34)$$

The maximum edge-weight clique of $P_{MEWC}(C'_{ex}, S'_{ex})$ is $\{v_{12}, v_{10}, v_7, v_4\}$, and its weight is 31, smaller than 35 of the inequality (34).

4.6 Vertex Renumbering

For the tightness of upper bounds, our algorithm renumbers vertex indexes before branch-and-bound. Let $\sigma(v) = \sum_{u \in N(v)} w_e(v, u)$. Our algorithm adopts following two policies:

Policy 1
    Vertices of large $\sigma(\cdot)$ are given large indexes.
Policy 2
    Vertices in an independent set are given consecutive indexes.

Similar policies are adopted by a branch-and-bound algorithm for MWCP [24]. Since $c[i] \leq c[i + 1]$ for any i, policy 1 is adopted to make $c[\cdot]$ small. Policy 2 also makes $c[\cdot]$ small. After renumbering, if $\{v_i, v_{i-1}, \ldots, v_1\}$ is an independent set, $c[i] = c[i - 1] = \cdots = c[1] = 0$.

According to these policies, our algorithm renumbers vertices by greedy vertex coloring as following.

(1) Maximize an independent set I_1 by adding vertices in V in nonincreasing order of $\sigma(\cdot)$. Give indexes $v_n, v_{n-1}, \ldots$ to vertices in order of addition.
(2) Maximize an independent set I_2 by adding vertices in $V \setminus I_1$ in nonincreasing order of $\sigma(\cdot)$. Give unused largest indexes to vertices in order of addition.
(3) Repeatedly maximize independent sets $I_3, I_4, \ldots, I_k$ and give unused indexes to vertices until $\cup_{i=1}^{k} I_i = V$.

5 Numerical Experiments

We implemented our algorithm EWCLIQUE by C++ and compare with previous methods in formulations of QP, IP and MIP solved by IBM mathematical programming solver CPLEX 12.5. The compiler is g++ 5.4.0 with optimization option -O2. The OS is Linux 4.4.0. The CPU is Intel®Core™i7-6700 CPU 3.40 GHz. RAM is 16 GB. Note that CPLEX is a multi-thread solver based on branch-and-cut, and our algorithm is a single-thread solver based on branch-and-bound. In the tables, d denotes the edge density that is $\frac{2|E|}{|V|(|V|-1)}$.

5.1 Random Graphs

We generated uniform random graphs as benchmarks. Edge-weights are integer values from 1 to 10. In each condition, we generated 10 instances and calculated the average computation time and average search tree size. The results for random graphs are shown in Table 1.

In all cases, our algorithm can solve them faster than CPLEX. For the instances that CPLEX needs several hundred seconds, our algorithm can solve them in less than a second. CPLEX cannot solve instances whose $|V|$ is greater than several hundred. Our algorithm solved $|V| = 15{,}000$ for sparse graphs. From the results, we confirmed that our algorithm is better than formulations solved by CPLEX.

5.2 DIMACS Benchmark Graphs

DIMACS is a benchmark set often used for MCP [29]. Since DIMACS graphs are not edge-weighted, we give edge-weights $w_e(v_i, v_j) = (i + j) \mod 200 + 1$, like the experiments in [13, 21].

The results for DIMACS are shown in Table 2, where each value is obtained by a single measurement. Instances that all algorithms cannot solve in 1000 s are not shown in the table. Excepting three instances: c-fat200-5, c-fat-500-10 and san200_0.7_1, our algorithm is fastest.

6 Conclusion

In this paper we propose a new exact algorithm EWCLIQUE for MEWCP. It is based on branch-and-bound. For each subproblem, EWCLIQUE considers three components to calculate the upper bound for the weights of feasible solutions. In the upper bound calculation, EWCLIQUE regards some edge-weights as pseudo vertex weights for vertices. EWCLIQUE uses two upper bound calculations for rest edge-weights and pseudo vertex weights. Upper bounds are obtained by merging them.

With some benchmarks, we compared our algorithm and some formulations solved by CPLEX. We confirmed our algorithm EWCLIQUE is faster than previous methods.

Table 1 Experimental results for random graphs

$|V|$	d	Optimal weight	Computation time (s)				Number of Search tree nodes			
			EWCLIQUE	MIP	IP	QP	EWCLIQUE	MIP	IP	QP
300	0.1	60.7	Less than 0.01	91.54	124.94	313.46	1835.1	164460.4	77165.7	193796.7
350	0.1	64.8	Less than 0.01	190.96	258.84	706.68	2721.0	264576.7	102579.4	308754.9
15,000	0.1	174.7	460.90	Over 1000	Over 1000	Over 1000	702255007.1	–	–	–
250	0.2	97.3	Less than 0.01	64.26	315.26	250.55	6412.8	365899.4	704421.5	438526.7
280	0.2	102.4	Less than 0.01	119.47	664.09	417.06	9862.9	502859.6	1243089.2	636879.1
5500	0.2	254.8	440.29	Over 1000	Over 1000	Out of memory	1537843711.0	–	–	–
200	0.3	150.0	Less than 0.01	39.16	470.07	154.74	14169.8	438264.9	1375256.2	923563.2
250	0.3	155.5	0.01	97.16	Over 1000	531.50	34844.2	823443.6	–	2301720.2
2500	0.3	332.8	459.05	Over 1000	Over 1000	Out of memory	1824084938.8	–	–	–
160	0.4	185.5	0.01	21.98	528.41	97.61	31813.3	379811.3	2058272.9	1563023.3
200	0.4	224.0	0.02	57.54	Over 1000	428.47	70701.9	869013.2	–	4983953.9
1400	0.4	444.3	758.22	Over 1000	Over 1000	Over 1000	2993314273.1	–	–	–
140	0.5	272.7	0.02	21.28	556.29	111.73	88105.7	507343.1	2935500.6	3028762.4
170	0.5	300.6	0.06	52.72	Over 1000	498.58	224608.1	1356597.9	–	9767768.2
750	0.5	560.3	603.91	Over 1000	Over 1000	Over 1000	2342210511.1	–	–	–

(continued)

Table 1 (continued)

$|V|$	d	Optimal weight	Computation time (s)				Number of Search tree nodes			
			EWCLIQUE	MIP	IP	QP	EWCLIQUE	MIP	IP	QP
120	0.6	399.0	0.05	18.10	405.38	129.01	208388.4	771892.2	3274026.2	4976548.2
130	0.6	424.6	0.07	28.57	634.81	256.59	288494.1	1183461.1	4496672.8	8570931.1
450	0.6	754.2	716.48	Over 1000	Over 1000	Over 1000	2725875895.6	–	–	–
100	0.7	583.5	0.11	15.12	262.75	125.42	437892.1	831713.2	3213925.0	6703943.1
110	0.7	607.1	0.24	31.23	557.70	356.28	950029.7	1657212	5078319.2	15772011.0
270	0.7	1049.7	589.15	Over 1000	Over 1000	Over 1000	2141882035.0	–	–	–
80	0.8	879.0	0.16	7.28	135.77	71.41	617626.3	517672.5	2107836.8	5237500.2
90	0.8	978.0	0.44	21.51	386.01	266.13	1578193.4	1294209.2	5293378.0	16189357.9
170	0.8	1580.2	485.50	Over 1000	Over 1000	Over 1000	1510657832.0	–	–	–
70	0.9	1708.4	0.62	3.80	50.80	16.69	2355972.7	258544.6	993629.0	1335799.1
80	0.9	2059.2	2.93	14.84	181.36	118.65	9974393.5	902841.7	2941909.6	7694204.9
110	0.9	2666.4	590.83	Over 1000	Over 1000	Over 1000	1951189872.0	–	–	–

Table 2 Experimental results for edge-weighted DIMACS

Graph	|V|	d	Optimal weight	Computation time (s)				Number of Search tree nodes			
				EWCLIQUE	MIP	IP	QP	EWCLIQUE	MIP	IP	QP
brock200_1	200	0.7454	21,230	338.31	Over 1000	Over 1000	Over 1000	1,328,614,116	–	–	–
brock200_2	200	0.4963	6542	0.10	109.66	Over 1000	Over 1000	345,371	3,241,394	–	–
brock200_3	200	0.6054	10,303	1.27	743.58	Over 1000	Over 1000	4,282,305	23,009,456	–	–
brock200_4	200	0.6577	13,967	4.84	Over 1000	Over 1000	Over 1000	13,814,425	–	–	–
c-fat200-1	200	0.0771	7734	Less than 0.01	4.80	4.92	62.62	632	1565	11,981	2951
c-fat200-2	200	0.1626	26,389	Less than 0.01	4.72	10.62	68.58	6780	3819	24,452	4263
c-fat200-5	200	0.4258	168,200	74.31	7.06	15.34	85.97	138,193,445	5010	22,597	2421
c-fat500-1	500	0.0357	10,738	Less than 0.01	171.91	51.33	749.86	1605	25,847	82,386	9931
c-fat500-2	500	0.0733	38,350	Less than 0.01	399.90	144.32	992.06	4679	95,599	158,428	17,113
c-fat500-5	500	0.1859	205,864	0.43	264.44	581.82	Over 1000	1,227,023	17,959	–	24,738
c-fat500-10	500	0.3738	804,000	Over 1000	745.93	Over 1000	Over 1000	–	486,168	–	–
DSJC500_5	500	0.5019	9626	44.43	Over 1000	Over 1000	Over 1000	200,152,687	–	–	–
hamming6-2	64	0.9048	32,736	Less than 0.01	0.07	3.70	0.23	896	4545	19,827	7158
hamming6-4	64	0.3492	396	Less than 0.01	0.22	2.68	0.66	340	11,550	18,194	7331
hamming8-2	256	0.9686	800,624	0.23	7.80	Over 1000	Over 1000	65,731	104,290	–	–
hamming8-4	256	0.6392	12,360	1.46	276.15	Over 1000	Over 1000	2,475,100	9,707,184	–	–
johnson8-2-4	28	0.5556	192	Less than 0.01	0.03	0.12	0.03	150	185	1922	827

(continued)

Table 2 (continued)

Graph	$|V|$	d	Optimal weight	Computation time (s)				Number of Search tree nodes			
				EWCLIQUE	MIP	IP	QP	EWCLIQUE	MIP	IP	QP
johnson8-4-4	70	0.7681	6552	Less than 0.01	0.40	8.74	2.93	3953	39,487	318,244	123,422
johnson16-2-4	120	0.7647	3808	0.25	57.4	Over 1000	Over 1000	1,905,154	4,543,549	–	–
keller4	171	0.6491	6745	0.70	167.84	Over 1000	Over 1000	2,158,496	8,809,323	–	–
MANN_a9	45	0.9273	5460	0.02	1.22	21.77	22.91	116,041	78,011	1,750,447	673,740
p_hat300-1	300	0.2438	3321	0.01	146.10	Over 1000	Over 1000	50,151	1,255,846	–	–
p_hat300-2	300	0.4889	31,564	42.90	Over 1000	Over 1000	Over 1000	134,486,327	–	–	–
p_hat500-1	500	0.2531	4764	0.13	Over 1000	Over 1000	Over 1000	468,371	–	–	–
p_hat700-1	700	0.2493	5185	0.52	Over 1000	Over 1000	Over 1000	1,678,557	–	–	–
p_hat1000-1	1000	0.2448	5436	2.92	Over 1000	Over 1000	Over 1000	9,890,185	–	–	–
p_hat1500-1	1500	0.2534	7135	32.73	Over 1000	Over 1000	Over 1000	106,284,583	–	–	–
san200_0.7_1	200	0.7000	45,295	54.88	28.72	Over 1000	Over 1000	387,149,894	662,303	–	–
san200_0.7_2	200	0.7000	15,073	17.86	Over 1000	Over 1000	Over 1000	48,732,878	–	–	–
san200_0.9_1	200	0.9000	242,710	12.56	206.01	Over 1000	Over 1000	12,731,307	5,158,955	–	–
san200_0.9_2	200	0.9000	178,468	833.49	Over 1000	Over 1000	Over 1000	303,169,816	–	–	–
san400_0.5_1	400	0.5000	7442	60.36	Over 1000	Over 1000	Over 1000	43,132,933	–	–	–
sanr200_0.7	200	0.6969	16,398	18.67	Over 1000	Over 1000	Over 1000	55,871,909	–	–	–
sanr400_0.5	400	0.5011	8298	9.04	Over 1000	Over 1000	Over 1000	36,003,126	–	–	–

References

1. Alidaee, B., Glover, F., Kochenberger, G., Wang, H.: Solving the maximum edge weight clique problem via unconstrained quadratic programming. Eur. J. Oper. Res. **181**(2), 592–597 (2007)
2. Aringhieri, R., Bruglieri, M., Cordone, R.: Optimal results and tight bounds for the maximum diversity problem. Found. Comput. Decis. Sci. **34**(2), 73 (2009)
3. Bahadur, K., Akutsu, T., Tomita, E., Seki, T.: Protein side-chain packing problem: a maximum edge-weight clique algorithmic approach. In: The Second Conference on Asia-Pacific Bioinformatics, vol. 29, pp. 191–200. Australian Computer Society, Inc. (2004)
4. Batsyn, M., Goldengorin, B., Maslov, E., Pardalos, P.M.: Improvements to MCS algorithm for the maximum clique problem. J. Comb. Optim. **27**(2), 397–416 (2014)
5. Bogdanova, G.T., Brouwer, A.E., Kapralov, S.N., Östergård, P.R.: Error-correcting codes over an alphabet of four elements. Des. Codes Cryptogr. **23**(3), 333–342 (2001)
6. Brown, J., Dukka Bahadur, K., Tomita, E., Akutsu, T.: Multiple methods for protein side chain packing using maximum weight cliques. Genome Inf. **17**(1), 3–12 (2006)
7. Brown, K.L.: Combinatorial auction test suite (CATS) (2000). http://www.cs.ubc.ca/~kevinlb/CATS/
8. Cavique, L.: A scalable algorithm for the market basket analysis. J. Retail. Consum. Serv. **14**(6), 400–407 (2007)
9. Corman, S.R., Kuhn, T., McPhee, R.D., Dooley, K.J.: Studying complex discursive systems. Hum. Commun. Res. **28**(2), 157–206 (2002)
10. Corman, S.R., et al.: Pajek datasets: reuters terror news network. http://vlado.fmf.uni-lj.si/pub/networks/data/CRA/terror.htm
11. Fang, Z., Li, C.M., Xu, K.: An exact algorithm based on MaxSAT reasoning for the maximum weight clique problem. J. Artif. Intell. Res. **55**, 799–833 (2016)
12. Gary, M.R., Johnson, D.S.: Computers and Intractability: A Guide to the Theory of NP-Completeness. W. H. Freeman and Company (1979)
13. Gouveia, L., Martins, P.: Solving the maximum edge-weight clique problem in sparse graphs with compact formulations. Eur. J. Comput. Optim. **3**(1), 1–30 (2015)
14. Horaud, R., Skordas, T.: Stereo correspondence through feature grouping and maximal cliques. IEEE Trans. Pattern Anal. Mach. Intell. **11**(11), 1168–1180 (1989)
15. Kc, D.B., Akutsu, T., Tomita, E., Seki, T., Fujiyama, A.: Point matching under non-uniform distortions and protein side chain packing based on efficient maximum clique algorithms. Genome Inf. **13**, 143–152 (2002)
16. Li, C.M., Jiang, H., Manyà, F.: On minimization of the number of branches in branch-and-bound algorithms for the maximum clique problem. Comput. Oper. Res. **84**, 1–15 (2017)
17. Martí, R., Gallego, M., Duarte, A.: A branch and bound algorithm for the maximum diversity problem. Eur. J. Oper. Res. **200**(1), 36–44 (2010)
18. McCreesh, C., Prosser, P., Simpson, K., Trimble, J.: On maximum weight clique algorithms, and how they are evaluated. In: International Conference on Principles and Practice of Constraint Programming, pp. 206–225. Springer (2017)
19. Östergård, P.R.: A new algorithm for the maximum-weight clique problem. Nordic J. Comput. **8**(4), 424–436 (2001)
20. Park, K., Lee, K., Park, S.: An extended formulation approach to the edge-weighted maximal clique problem. Eur. J. Oper. Res. **95**(3), 671–682 (1996)
21. Pullan, W.: Approximating the maximum vertex/edge weighted clique using local search. J. Heuristics **14**(2), 117–134 (2008)
22. San Segundo, P., Rodríguez-Losada, D., Jiménez, A.: An exact bit-parallel algorithm for the maximum clique problem. Comput. Oper. Res. **38**(2), 571–581 (2011)
23. Shimizu, S., Yamaguchi, K., Masuda, S.: Mathematical programming formulation for the maximum edge-weight clique problem. IEICE Trans. Fundam. Electron. Commun. Comput. Sci. (in Japanese) **J100-A**(8), 313–315 (2017)
24. Shimizu, S., Yamaguchi, K., Saitoh, T., Masuda, S.: Fast maximum weight clique extraction algorithm: optimal tables for branch-and-bound. Descr. Appl. Math. **223**, 120–134 (2017)

25. Sørensen, M.M.: New facets and a branch-and-cut algorithm for the weighted clique problem. Eur. J. Oper. Res. **154**(1), 57–70 (2004)
26. Sorour, S., Valaee, S.: Minimum broadcast decoding delay for generalized instantly decodable network coding. In: Global Telecommunications Conference (GLOBECOM 2010), pp. 1–5. IEEE (2010)
27. Tomita, E., Kameda, T.: An efficient branch-and-bound algorithm for finding a maximum clique with computational experiments. J. Glob. Optim. **37**(1), 95–111 (2007)
28. Tomita, E., Yoshida, K., Hatta, T., Nagao, A., Ito, H., Wakatsuki, M.: A much faster branch-and-bound algorithm for finding a maximum clique. In: International Workshop on Frontiers in Algorithmics, pp. 215–226. Springer (2016)
29. Trick, M., Chvatal, V., Cook, B., Johnson, D., McGeoch, C., Tarjan, B., et al.: DIMACS implementation challenges. http://dimacs.rutgers.edu/Challenges/
30. Yamaguchi, K., Masuda, S.: A new exact algorithm for the maximum weight clique problem. In: 23rd International Conference on Circuits/Systems, Computers and Communications (ITC-CSCC08), pp. 317–320 (2008)

A Software Model for Precision Agriculture Framework Based on Smart Farming System and Application of IoT Gateway

Symphorien Karl Yoki Donzia, Haeng-Kon Kim and Ha Jin Hwang

Abstract Contemporary society is seriously threatened with food as part of the world due to the continuous increase in world population, the degradation and decline of agricultural lands due to high industrialization, climate change and the aging of the population. Therefore, modern society is studying different solutions to solve human food. In this paper, a framework for precision agriculture using IoT Gateway is proposed for solving human food, and the productivity of crops must be increased first. IoT solution through architecture, platforms and IoT standards, or the use of interoperable IoT technologies beyond the adopters in particular, simplifying existing proposals. Connecting different sensors, connected devices, developing intelligent breeding systems as much as possible. One of our aims is to manage and challenges. We provide a techniques and technologies applications during our work. The result shows that the advantages of various types of sensors for agriculture services in their decision making. And a proposed architecture for Agriculture Mobile services based on Sensor Cloud substructure that helps farms and IoT applications are effective in intelligent farming system.

Keywords Farming · Cloud · IoT · Wi-Fi · Agriculture · Gateway

1 Introduction

In the current agricultural scenario, farmers face problems of crop pests/diseases, water shortages, weak linkages with the food supply chain, transportation, and the supply chain of food and agro-food products. Who has to focus on the current world

S. K. Y. Donzia · H.-K. Kim (✉)
Daegu Catholic University, Gyeongsan, South Korea
e-mail: hangkon@cu.ac.kr

S. K. Y. Donzia
e-mail: yoki10@cu.ac.kr

H. J. Hwang
Sunway University Business School, Sunway University, Subang Jaya, Malaysia
e-mail: hjwang@sunway.edu.my

© Springer International Publishing AG, part of Springer Nature 2019
R. Lee (ed.), *Computational Science/Intelligence & Applied Informatics*, Studies in Computational Intelligence 787, https://doi.org/10.1007/978-3-319-96806-3_4

situation is; the loss of agricultural land due to the rapid growth of the population. As the population increases, more land is needed for housing that tends to lose land for agricultural purposes, but to live, the same population also needs food [1]. To feed the rapidly growing world population in the coming years, agriculture must produce more. Computer technology in the last decade has changed a lot in many areas. When observing the evolution of the computer, in the early ages, the era of the mainframe has experienced a great evolution in which only an expert has accessed the computer. But today, popular devices such as laptops, tablets and smartphones are maintained and accessible to all ages. The third wave of computing is already in us, resulting in the departure of central and personal computers [2]. Technologies such as satellite navigation, the sensor network, network computing, contextual calculation and the omnipresence of information technology and decision making [3]. Agriculture also benefits from the technological innovations that support quantitative and qualitative food production. The omnipresent computer science in agriculture emerges remarkably in this environment of ubiquitous and fast processing, thanks to the network of wireless sensors (WSN). Internet of Things (IoT) has the ability to transform world life into more efficient industries, connected cars and smart cities, all components of the Internet of Things equation. However, the application of technologies such as IoT in agriculture could have the greatest impact. Precision agriculture is one of the famous IoT implementation in the agricultural area or many institution are utilizing the methode anywhere in the world. Crop Metric is a exactitude agriculture institution focused on cutting-edge agronomic compound.

In fact, we have three types of agricultural drones: evaluation of crop health, irrigation, crop monitoring, crop fumigation, planting and soil and field analysis. And we also have cattle monitoring that reduces labor costs because farmers can locate their livestock using IoT sensors and the third is Smart Greenhouses which smart greenhouse can be create with IO; Monitors and command the weather, remove the need for by hand intervention.

These models are interconnected and designed to facilitate the flow of data from one to the other.

Climate data processing tools: statistical downscaling and interpolation tools for data preparation for hydrological modeling and culture tools.

Crop models: simulate crop growth in climate change scenarios, using data produced by climate data processing tools. Hydrologic model: models watershed hydrology under climate change scenarios, using data produced by climate data processing tools. Economic model: simulates the impact of climate change-induced yield changes in national economies [4].

Current Issue: As the economy grows, the demand for quality of life and material goods increases, and the group of low-income people can gradually respond to the demands of younger generations in Africa. The increase of incentives and employment opportunities in the non-agricultural sector also leads to a diversification of the agricultural labor force. The problem of the aging of farmers has also become important, which adds to the weakening of the quality and quantity of human resources. The lack of agricultural machinery still exists. In fact, the problem of "human hunting" in the agricultural sector of the non-agricultural sector becomes serious. However, in many developed countries, such as Taiwan, Japan and Korea, thanks to the development of industry and companies, the improvement of the standard of living has led to a greater demand for public, domestic and industrial water. Once again, due to the difficulties in developing new water sources, as well as the problems of environmental protest, the demand for the transfer of agricultural water resources to non-agricultural sectors has increase.

Our Goal: our goals are to make life easier for people, and most of the connectivity scenarios that farmers use are things like Satellite's customized cellular solutions and that's why connectivity is too much. The key is to use a new technology called Wi-Fi developed in Microsoft Research, which consisted of using unused TV spectrum for Wi-Fi connectivity to send Wi-Fi signals to the TV spectrum. The key was that, although in the cities there are many tricks of TW in farms that are very far from the cities, they are far from the TV towers, so there is a lot of spectrum. You can connect sensors, cameras and send drone videos to the cloud at high speed. Therefore, it is an important thing that the agricultural rhythms would implement in the carnation. For example, if we collect data from farms.

Our Motivation: the connectivity in the farms is weak, we could not really send all the data to the cloud. The reason why IoT in the Internet of Things is difficult to apply to agriculture is probably one of the most difficult problems.

2 Framework

2.1 Objectives of IoT Platform in Farm

We are looking at this technique which was developed around a couple of decades ago called precision farming the idea is that instead of the farmer thinking of the entire farm as a homogeneous piece of land example of absorb moisture differently some parts of the farm have are more fertile than the other we can plant seeds closer together rather than applying irrigation throughout the farm in a uniform manner apply only where it is needed and how much it is needed. So we think technology can get us there to enabling precision agriculture the reason though even the precision agriculture was developed a couple decades ago which has no really taken off and the biggest reason is that getting data from the farm is still very expensive, that is though these sensors exist deploying these sensors are non-trivia.

2.2 Objectives of IoT Platform in Farm

We target the following goals; Availability that the platform should have negligible down time. And the Capacity which should support sensors with widely varying requirement; pH sensor reporting few bytes of data to drones sending gigabytes of video. And also a Cloud Connectivity that several farming application, such as crop cycle prediction, seeding suggestions, farming practice advisory, etc. [5]. And the last one is Data Freshness what which a state sensor data from the farm can make application suggest incorrect courses of action to the farmer.

2.3 Design Decision

An overview of the system is given in Fig. 1. Here, we discuss the main design decisions. Sensors, drones and cameras are not normally compatible with TVWS. Therefore, to maintain compatibility with sensors with long-range high-bandwidth connectivity set up a two-layer hybrid network. In the second layer, the IoT Base Station provides a Wi-Fi interface for sensor connections and other devices. The Wi-Fi interface ensures that the farmer can not only connect most agricultural sensors, cameras and drones available on the market; but they can also use their phones to access agricultural productivity applications. Farm allows farmers to switch the Wi-Fi access base station for productivity applications, while on the farm. Given the low Internet connectivity on the farm, a naive approach to bringing all the data into the cloud is not working. We make the key observation that the data requirements of agricultural applications can be categorized into two broad categories: the immediate detailed data and the long-term summary data. This categorization allows an IoT gateway-based design for Farm. Gateway is detect in the farmer's house and

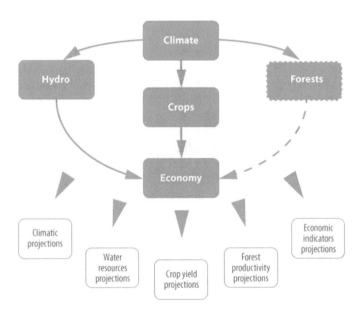

Fig. 1 Modeling system for agriculture

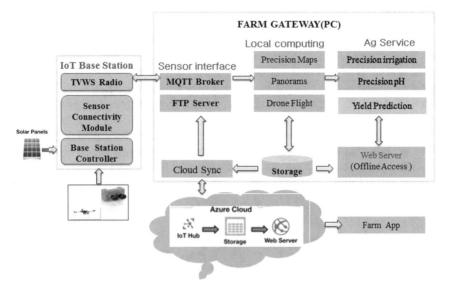

Fig. 2 Farm system overview

the area link or work for two functions: (a) produce summaries for future use and dispatch them to the cloud, (b) proposal applications that could be assuming locally. Summaries are respective detective of magnitude lower than farm raw data (Fig. 2).

IoT Platform as SaaS

As mentioned earlier, today a number of cloud providers provide IoT capable software as a service which acts as a platform to connect hundreds of smart sensors and make logical control over them. These are SAAS platforms where IoT platforms serve as the bridge for direct connection among spatially distributed devices and data storage centers.

3 Agriculture

The agricultural system includes the following elements: Sensors and drones. The farm uses ready-to-use sensors for its applications. Each sensor measures specific farm characteristics, such as soil moisture and soil pH, and transmits this data to the IoT base station over a Wi-Fi connection. In addition to the ground sensors, Farm takes Charge the cameras to monitor farms and drones [5]. The cameras are connected to the IoT base station via Ethernet or report data via Wi-Fi. They take periodic snapshots and transmit this data to the IoT base station. UAV flights are scheduled periodically or manually using the Farm app on the farmer's phone. IoT base station: the farm's IoT base station is powered by solar panels, supported by batteries and has three components.

3.1 IoT Gateway

The goal of the IoT gateway is to enable local services and create summaries from existing data to be sent to the cloud. We use a PC form factor device as the Farm gateway, which is typically placed in the farmer's house or office, whichever has Internet access. The gateway provides an interface for applications to run and create summaries to be sent to the cloud as well as to post data to the local web server. Furthermore, it includes a web service for the farmer to access detailed data when they are on the farm network [6]. First, the Farm gateway implements a web service, providing unique services that are different from the Farm web service in the cloud. Second, the gateway can operate offline, and still offer the most important services.

3.2 Precision Farming

The Techniques were present in earlier decades. However, largely take in with some innovative farmer [7]. To expand the theory of precision agriculture to smart agriculture which a farm show to advantage a smart network of interoperable agricultural target. The farm management position which addresses the seamless standardization of real-time trail. Smart analysis or smart command of applicable agricultural process operation. Other issues are to improve ease of use, affordability and adoption by simplifying existing solutions and involving communities of mainstream farmers besides early adopters.

3.3 Concept of Cloud Service

The goal of cloud computing is to develop a collective, well-maintained, resource scalable and easily deployable solution which would ease the deployment time for applications, thereby giving the flexibility for developers to focus only on their application. For concurrently running different enterprise applications on a single cloud instance, cloud providers leverage on virtualization technology. Virtualization segregates the disk memory into many virtual devices so that each virtualized server can co-exist with others and run their own application. Operating System virtualization makes the best use of computing resources and solves platform dependency. Additionally, memory allocation to the virtualized application can be dynamically allocated on demand, thereby, featuring scalable solution for the software. Cloud Service. Deploying a software package is a collective effort of platform support. These can be categorized as the layered stack [8]. (1) Data Center: Consists of the data center where computer storages are physically installed along with ventilation and security features. (2) Computer Network: Deploying interconnectivity between all the computers such that they can operate as a network. (3) Servers: A highly elastic and reliable server which can run 24 * 7 with load balancing features.

(4) Virtualization: This defines and provides separate disk partition within a computer to dynamically allocate resources to the deployed software. (5) Operating System, Middleware, and Run-Time Environment: Any software has its dependencies defined by runtime environment and operating system used. Due to this appropriate dependency need to be installed on the virtualized disk. (6) Application and Data: This refers to user-defined code which manages requests received from the server. Different applications may share all the above-mentioned features but differ in this layer based on their application.

4 Result and Evaluation

4.1 UAV Flight Planning

To know the affect of area coverage algorithms on drone flight time, we collate production of Farm in awning a considering area. As shown in Fig. 3, algorithm lead to sweeping sample from middle to the west. However, we compare the time taken to complete flights planned by the two algorithms to cover a given area. The maximum speed was set to 10 m/s and the altitude was set to 20 m. Figure down plots the time taken to complete a flight with the two algorithms in different area.

Finally, we evaluate the impact of our yaw control algorithm under different wind conditions. The maximum speed was set at 10 m/s and the altitude was set at 30 m. For each flight, we fully charge the battery. We measure the percentage of time saved by the Farm yaw control algorithm for each flight and plot it in the following figure.

Fig. 3 Deploy on the map

As shown in the figure, Farm can save up to 5% of the time depending on the wind speed. Also as the north-south wind component increases.

4.2 IoT Gateway Implementation

By software: collect, process and analyze data regularly to provide a broad vision to farmers in an accessible way. This software will use data from hardware sources (sensors), purchased by the farmer or by hardware companies that the software service provider associates with machinery or data manufacturers provided by third-party farmers or data service providers. public goods such as local governments. Based on the software, the presentation and analysis of the data differ. But now, most programs are available through computers, tablets and smartphones and usually include a dashboard of different customizable data sets for anyone. The software also helps farmers make critical decisions, some of which are listed in Table 1, allows farmers to save money in areas that do not need it and also improves yield [9] (Fig. 4).

4.3 Assessment

Agriculture is one of the most important aspects of human civilization. The uses of information and communication technologies (ICT) have made a significant contribution to the region in the last two decades. To estimate the proposed algorithm, we are using two inputs, involve the predefined IoT management action. In our work, an accurate which rice is a semiaquatic factory which prefers to expand in immersion [7]. The result of the algorithm is an IoT management process capable of detecting the risks related to OI in the agricultural supply chain, including perception, network and layer of application risks, which are summarized in this document. The ordered weighted average operator is used to quantitatively assess and classify these risks.

Table 1 IoT platforms

IoT software platform	Interaction	Security	Protocols for data collection
IoT analytics platform	Salesforce	SSL	mqtt
AWS IoT platform	REST API	SSL	MQTT, HRRP
IoT smart product platform		Link encryption	MQTT
IBM IoT foundation of farm cloud	REST AND real time	SSL	Websokets/MQTT
M2M application platform	REST API	SSL	MQTT

Fig. 4 Cloud based software architecture in conjunction with IoT gateway

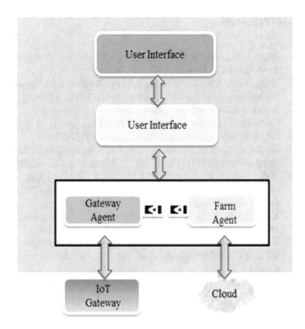

5 Conclusion

Future work will be developed with cloud-based frameworks for agriculture that also include IoT (Internet of Things) to work together with the fast-growing world and the massive increase in IoT. The experimental results indicate that the proposed approaches can be applied effectively to estimate the IoT Farm platform, although they are simple and effective. The experimental results show that the introduction of the characteristics of the drones can significantly increase the accuracy of the classification of the farm worker in the cloud. Studies on sensor technologies have been mentioned to provide an additional improvement to the system developed in the future. The development of wireless detection technologies such as Wi-Fi, wireless surveillance sensors makes the farm more convenient, economical and easy to install. We believe that the large IoT gateway and metadata play a crucial role in ubiquitous agriculture in the management of the increasingly complex amount of data associated with applications or services.

Acknowledgements This research was supported by the MSIP (Ministry of Science, ICT and Future Planning), Korea, under the ITRC (Information Technology Research Center) support program (IITP-2018-2013-1-00877) supervised by the IITP (Institute for Information & communications Technology Promotion).

References

1. Jagyasi, B., Mohite, J., Pappula, S.: Applications of mobile sensing technologies in precision agriculture. CSI Commun. 21–23 (2013)
2. Yang, J., Rao, B., Zimmermann, H.D.: Preface to the focus theme section 'pervasive computing/ambient intelligence'. Electron. Mark. **15**(1), 3 (2005)
3. Aqeel-ur-Rehman, Z.A.: Shaikh, Smart Agriculture, Application of Modern High Performance Networks. Bentham Science Publishers Ltd., pp. 120–129 (2009)
4. http://www.fao.org/climatechange/mosaicc/66705/e
5. Sivamani, S., Bae, N., Cho, Y.: A smart service model based on ubiquitous sensor networks using vertical farm ontology. Int. J. Distrib. Sens. Netw. **2013** (2013)
6. Wei, J., Zhao, Y., Jiang, K., Xie, R., Jin, Y.: Analysis farm: a cloud-based scalable aggregation and query platform for network log analysis. In: 2011 International Conference on Cloud and Service Computing (CSC), pp. 354–359, Dec 2011
7. Król, D., Kitowski, J.: Self-scalable services in service oriented software for cost-effective data farming. Future Gener. Comput. Syst. **54**, 1–15 (2016)
8. Zhu, Q., et al.: IoT gateway: bridging wireless sensor networks into internet of things. In: 2010 IEEE/IFIP 8th International Conference on Embedded and Ubiquitous Computing (EUC). IEEE (2010)
9. Pearson, S.: Taking account of privacy when designing cloud computing services. In: ICSE Workshop on Software Engineering Challenges of Cloud Computing, CLOUD'09, pp. 44–52. IEEE (2009)

Components of Mobile Integration in Social Business and E-commerce Application

Mechelle Grace Zaragoza, Haeng-Kon Kim and Youn Ky Chung

Abstract E-commerce and Social Business solutions are developed through the integration of different components. Distributed software agents are very promising in developing an increasingly penetrating component and middleware technology. Agents are specialized types of components that offer greater flexibility than traditional components. This study focuses on the development of a software agent that could be used to assemble different types of frameworks written and constructed by different developers from different platforms. This paper proposed an application of mobile integration components in social enterprise and e-commerce as a software development methodology to simply integrate different basic components of the technology into a single web-based solution. We propose a systematic development process for the software agent using components and UML. We first developed the agent component specification and modeled it. Based on this, we developed a mobile application for social business applications. We integrate the module-based software framework into Drupal's content management system.

Keywords E-commerce/social business · Web-based solution · Component based

1 Introduction

E-commerce and Social Business solution consists of built-in components that work differently but have a purpose. From this perspective, this paper proposed a mobile integration component in social enterprises and the application of e-commerce as a

M. G. Zaragoza · H.-K. Kim (✉)
Daegu Catholic University, Daegu, South Korea
e-mail: hangkon@cu.ac.kr

M. G. Zaragoza
e-mail: mechellezaragoza@gmail.com

Y. K. Chung
Kyungil University, Daegu, Korea
e-mail: ykchung@kiu.ac.kr

© Springer International Publishing AG, part of Springer Nature 2019
R. Lee (ed.), *Computational Science/Intelligence & Applied Informatics*, Studies in Computational Intelligence 787, https://doi.org/10.1007/978-3-319-96806-3_5

software development methodology to simply integrate the different components of technology into a single solution based on the Web.

Social collaboration is an integrated set of tools that enable real-time knowledge exchange, greater productivity, and faster innovation. The purpose of this study is to design a framework for the application of social enterprises that will apply a component-based development methodology. Nowadays, there is a need for rapid development of the mobile application independent of the platform. Flexible design of Social Business application systems for mobile is required. The objectives of this study are to develop a framework with a component-based development methodology, design a social enterprise application as a case study, and develop a mobile application as a result of this study. The most effective approach to enabling a Social Business solution to help people discover their experience, develop social networks, and leverage relationships. A social enterprise makes it easier for your employees and customers to find the information and experience they are looking for.

It helps groups of people come together in communities of common interest and coordinate their efforts to achieve better business results faster. Encourages, supports and leverages innovation and idea creation and relies on the intelligence of the crowd. The contributions of this paper are the development of the framework using a component-based development methodology for faster integration of software independent of the platform. This study designs a new social business application model using a content management system, Drupal, and the result of this study is the integration and development of the mobile application.

2 Background of the Study

2.1 Software Reuse

The reuse of software allows the development of software of better quality and at a lower cost. Software reuse environments are seeking to improve the reuse of software artifacts, especially when performed at the beginning of the software life cycle [1]. The reuse of software has been practiced from the programming started. Reuse as a separate study field in the software Engineering, however, is often attributed to Doug Mcilroy paper that proposed to found the software industry in reusable components. The introduction by the neighbors of the concepts of domain and domain analysis. Active research areas of reuse in the last twenty years include reuse libraries, domain engineering methods and tools, reuse design, design templates, domain specific software architecture, components, generators, measurement and experimentation, and business and finance. The ideas that emerge from this period include systematic reuse, reuse design principles like the three C model, module interconnection languages, analysis of similarity/variability, point of variation, and various approaches to the domain specific generators [2].

Relational and object-oriented paradigms are simply too different to be harmonized, leading to a series of difficulties known as impedance mismatch [3]. To explicitly support reuse-enabling guidelines, the model has four distinct process elements:

Create. This process provides reusable resources appropriate for the process of use. Resources can be new, reworked or purchased, and of various types, such as code, interfaces, architectures, tests and tools. The activities include analysis of inventory and domain of existing applications and assets, evaluation of user needs, market trends and technology, as well as architecture and definition and construction of components. This process reuses assets to produce products (applications or system). The activities include the review of the domain and asset models, analysis of product requirements, adaptation of assets, product development and specification of proposed components.

Support. This process supports the process of global reuse, administration and maintenance of the asset collection. Activities include certification of new assets, classification for library storage, correspondence of use needs with assets, provision of user support and the collection of comment reports.

Manage. This process restricts and controls the other elements. You must plan, launch, manage, track, prioritize, coordinate and improve the reuse process. Activities include establishing priorities and schedules to build new assets, resolve conflicts when resources are not available, establish training and establish direction.

Coordination. Can take place at various levels, within each project of creation and use, through a set of users, and even the management of a complete system of "producer-consumer" projects [4].

2.2 Component Based Development

Component-Based Development claims to offer a radically new approach to the design, construction, implementation and evolution of software applications. Software applications are assembled from components from a variety of sources; the components themselves may be written in several different programming languages and run on several different platforms. CBD architecture is being used nowadays and the research on how making it more efficient is the focus of this study. A component re-used is one of the most convenient ways for the fast software production. There have been many methods on how to do this and it does involve more technical and detailed view. In this paper we tried to integrate the concept of CBD to develop a mobile enterprise application. We believed that enterprise application uses software components that are being re-used repeatedly; hence, component re-used for mass application developments is necessary [5, 6]. Component Based Development (CBD) is a popular methodology to develop a mobile component through component re-used. One of the interesting researches is the enterprise mobile application development with CBD.

2.3   Social Business Systems

Social business, as the term has been commonly used since, was defined by Nobel Peace Prize laureate Prof. Muhammad Yunus and is described in his books, creating a world without poverty [7].

A social business is a company created with the sole purpose of solving a social problem in a financially self-sustainable way. A good social business combines an unwavering focus on meeting social needs with entrepreneurial energy, market discipline, and great potential for replicating and scaling successful enterprises.

As the rapid growth of social networking and mobility has erased some of the boundaries that separated individuals in the past, people increasingly use their relationships with other people to discover and use information to accomplish innumerable tasks. New opportunities for growth, innovation and productivity exist for organizations that encourage people—employees, customers and partners—to engage and build trusted relationships. Individuals are using social networking tools in their personal lives, and many are also incorporating it into their work lives—regardless of whether it's sanctioned by their employers. Astute organizations will embrace social software and find the most effective ways to utilize it to drive growth, improve client satisfaction and empower employees.

Figure 1 show the Social Business use case centric model based on IBM architecture. It has four components, the social collaboration, social analytics, and social content and social user experience. There four components should be met for a successful social business solution. Integrating these components requires a different type of technology integration.

There are four main Platforms for Social Business. These are social networking, social analytics, social content, and social user experience [8].

Social networking platforms should meet some considerations

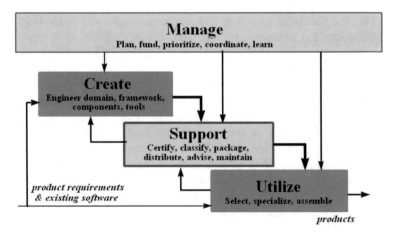

Fig. 1 Basic reuse process model

Social Analytics

- Infused into social platforms
- Leverage social data to under hidden relationships
- Make determinations on what people think and might do
- Integrated solutions

Social Networking

- People-centric, relationship driven
- Openness
- Transparent work and open decision making
- Connected and discoverable
- Business driven
- Adaptable

Social Content

- User Contributed
- Co-creation
- Developing content for web, mobile, and social channels
- Engaging (Fig. 2)

Social User Experience

- Role-based, relationship driven social, web, and mobile experiences
- Dynamic, adaptable, and personal
- Engaging

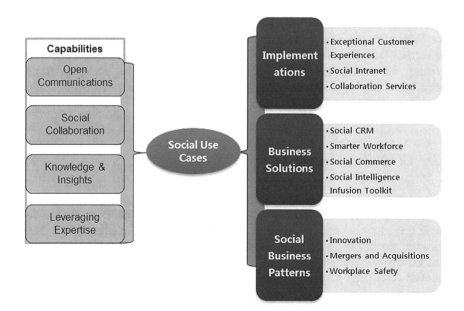

Fig. 2 Social business use case centric model

2.4   Business and Finance

The ultimate goal of domain and systematic engineering the reuse of the software is about improving the quality of the products and the services that a company provides and, therefore, maximize profits It is easy to lose sight of this objective when given the technical challenges of software reuse and however, the reuse of the software will only be successful if commercial sense Capital can be spent by an organization many ways to maximize the return of shareholders. Software the reuse will be chosen only if it is possible to make a good alternative option to use capital. Business-related reuse research has identified. Structure to support corporate reuse programs, organized process models to adopt reuse and models to estimate return on investment of a reuse program [4].

3   Social Business System Modeling

3.1   Component Identification

The component identification stage takes as input the business concept model and the use case model of the requirements workflow. It assumes an application overlay that includes a separation of system components and commercial components. Its objective is to identify an initial set of business interfaces for the business components and an initial set of system interfaces for the system components, and to bring them together in the initial architecture of the components. The business-type model is an intermediate artifact from which the initial business interfaces are formed. It is also used later, at the component specification stage, as a raw material for the development of interface information models [9]. You should also consider all existing components or other software assets, as well as the architectural models you plan to use. At this point, it is a fairly broad substance, destined to define the panorama of the components and interfaces for further refinement.

3.2   Component Interaction

The component interaction step examines how each of the system operations will be performed using the architecture of the component. Use interaction models to discover operations in business interfaces. As more and more interactions are planned, common operations and usage patterns emerge that can be factored and reused. Responsibility options become clearer and operations moed from one interface to another. Alternative groups of interfaces for the components can be studied. It is time to think about managing the references between the component objects so that the dependencies are minimized and the referential integrity policies are taken into account. The component interaction stage is the stage where all the details of the structure of the system appear, with a clear understanding of the dependencies between the components, up to the level of individual operation.

3.3 Component Specification

The last stage of the specification of operations and restrictions takes place. For a given interface, this means defining the potential states of the component objects in an interface information model, then specifying the prerequisites and after the operation, and capturing the business rules as constraints. These interface specification details bear witness to the specification of constraints that are specific to a particular component specification and independent of each interface. These component specification constraints determine how the type definitions in the individual interfaces will match each other in the context of the component. Architecture should not change materially at this stage. These detailed specification tasks should be carried out once the architecture is stable and all interface operations have been identified. Writing the precise rules for each operation can help you discover missing parameters or missing information, but the focus is on detail details in a stable frame (Fig. 3).

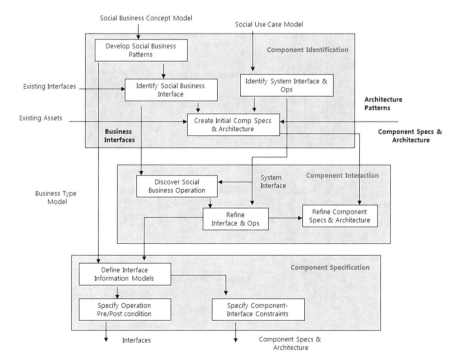

Fig. 3 Social business specification workflow

4   Mobile Integration in Social Business and E-commerce Application

4.1   UML Modeling Technique

The social business concept template diagram is a class diagram that describes the business concept template. An interface specification diagram describes the interface specification. And so on, with the diagram of the business type model, the component specification diagrams and the component architecture diagram, each representing the corresponding artifacts.

Figures 4 and 5 shows the architecture of the components. It is a collection of application-level software components, their structural relationships and their behavioral dependencies. A component architecture can be applied to a single application or in a broader context, such as a set of applications that serve a particular social enterprise process domain.

4.2   Implementation

The following are needed software and steps to develop the mobile application:

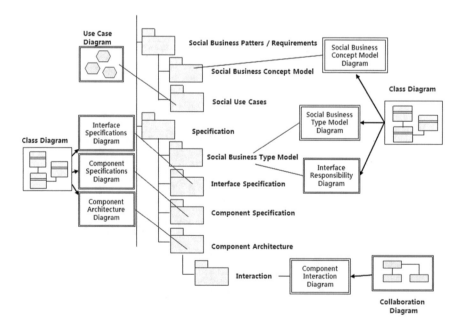

Fig. 4 Social business component modeling diagram

Fig. 5 Component
architecture

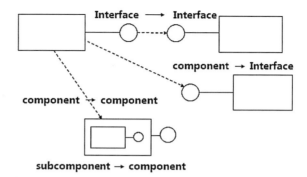

1. Install Web Server (xampp)
2. Install drupal
3. Customize the installed Drupal based on web layout and contents
4. Work on the needed modules for social business application [10, 11].

5   Conclusion

This study is focusing on developing a software agent that could be used to assemble different type of frameworks which were written and built by different developers of different platforms. This paper proposed a Component of Mobile Integration in Social Business and E-commerce Application as a software development methodology to simply integrate the different technology building blocks into one web-based solution. We first developed the agent component specification and modeled it. Based on this, we developed a mobile application for social business application as a case study. We integrate the developed software framework as a module in Drupal content management System as our case study.

Acknowledgements This Research was partially supported by KyungIl University, Republic of Korea.

References

1. Mahmood, S., Ahmed, M., Alshayeb, M.: Analysis and evaluation of software artifact reuse environments. Int. J. Softw. Innov. (IJSI) **2**(2), 54–65 (2014)
2. Frakes, W.B., Kang, K.: Software reuse research: Status and future. IEEE Trans. Softw. Eng. **31**(7), 529–536 (2005)
3. Pereira, Ó.M., Aguiar, R.L., Santos, M.Y.: Reusable business tier components: based on CLI and driven by a single wide typed service. Int. J. Softw. Innov. (IJSI) **2**(1), 37–60 (2014)
4. Griss, M.L.: Systematic software reuse: architecture, process and organization are crucial. Fus. Newsl. (1996)

5. Agner, L.T.W., Soares, I.W., Stadzisz, P.C., Simao, J.M.: Model refinement in the model driven development context. J. Comput. Sci. **8**(8), 1205–1211 (2012). ISSN 1549-3636
6. Cheesman, J., Daniels, J.: UML Components: A Simple Process for Specifying Component-Based Software. The Addison-Wesley Object Technology Series 2000. ISBN: 0-201-70851-5
7. Yunus, M.: Creating a World without Poverty: Social Business and the Future of Capitalism. Public Affairs, p. 320 (2009). ISBN 978-1-58648-667-9
8. Social Business Architecture, IBM Corporation 2014, Accessed on 10 March, 2015
9. Cheesman, J., Daniels, J.: UML Components, A simple Process for Specifying Component-Based Software. The Addison-Wesley Object Technology Series (2001)
10. Kim, H.-K.: Mobile agent development with CBD on ABCD architectures. In: International MultiConference of Engineers and Computer Scientists (2013)
11. Kim, H.-K.: Design of web-based social business solution architecture based on CBD. Int. J. Softw. Eng. Appl. **9**(2), 271–278 (2015)

Design and Evaluation of a MMO Game Server

Youngsik Kim and Ki-Nam Kim

Abstract Many large-scale online genre games such as Massive Multi-player Online Role Playing Game (MMORPG) are attracting attention in the game market. In a game server connected to hundreds or thousands of users, a large number of packets come and go between the server and the client in real time. For the server to endure these loads, IOCP (Input/Output Completion Port) and multi-thread are necessary. This paper implements a simple MMO Game Server using IOCP and evaluates its performance. Also, IOCP packet design and processing method are presented. The Simple MMO Game Server implemented in this paper also supports multi-thread synchronization and dead reckoning.

Keywords Massive multi-player online role playing game (MMORPG) · Game server · Multi-thread · IOCP (Input/output completion port)

1 Introduction

MMOFPS (Massive Multi-player Online First Person Shooting) game battleground in Fig. 1, which has 100 users living on a large island in Korea recently). The Battle Ground, which was released at Blue Hall, had the shortest record of early access to the steam platform with sales of one million units in 16 days. Recently, the development of online games that are getting bigger than in the past is active. In the past, the development of large-scale games in more than 100 people has been a trend these days, compared to the 8-person, 16-person online games. In a large online game, one server is responsible for as few as a hundred and as many as a thousand clients. The more clients connected to the server, the more network loads are placed on the server.

Y. Kim (✉) · K.-N. Kim
Department of Game and Multimedia Engineering, Korea Polytechnic University,
Siheung, Republic of Korea
e-mail: kys@kpu.ac.kr

K.-N. Kim
e-mail: ddous@kpu.ac.kr

© Springer International Publishing AG, part of Springer Nature 2019
R. Lee (ed.), *Computational Science/Intelligence & Applied Informatics*, Studies
in Computational Intelligence 787, https://doi.org/10.1007/978-3-319-96806-3_6

Fig. 1 Screen shot of PLAYERUNKNOWN'S BATTLEGROUND2 related work

In recent years, as almost everyone is using the Internet, there are also users who play online games. MMORPG (Massive Multi-player Online Role Playing Game) is the trend of the game which is recently released on mobile, and large-scale online games are also being released on PC and console side.

To design a large-scale online game server, IOCP (Input/Output Completion Port) [1] is almost essential in a Windows environment. This means that you will specify the port that will handle the completion of input and output. Since the notification at the completion time of input and output is processed by the overlapped input/output, this technique can be regarded as an extension of the superposition input/output technique of the window. A port is an object created to handle a task or service. It will be easy to recall that the port of a socket is an object for transferring data I/O to a particular service. Once you understand the characteristics of these ports, you can think about how to process the data by creating a port dedicated to notification at the completion of input and output. When I/O is completed from an input/output device (here, it is limited to a socket), this completion report is accumulated in the input/output completion queue. The thread wakes up, and the thread reads the completion report in the queue to process the data.

The advantage of this approach is that you can create a thread in advance and wait for I/O completion report, as opposed to the multithreaded way of creating a thread every time the client connects. By creating such a reasonable number of worker threads, the operating system can wake up idle worker threads. This approach is similar to a thread or process pool, but there is a difference. In general, it is not easy to allocate tasks to idle threads in the thread pool approach. To do this, I need to use some tricky techniques, and the IOCP does not need to worry about the developer, but the operating system picks up and wakes up the thread.

Multithreaded programming is inevitable to use IOCP. Multi-threading is not essential, but multi-threading is essential to building a game server using IOCP. Much work has been done to use multithreaded environment because synchronization is essential and it is the most important factor to improve performance. In this paper, we propose a basic server architecture and design to implement MMORPG and provide the following three techniques. First, we design and design the packet to be used in the MMORPG server. Second, we propose a synchronization technique for prevention of deadlock and data race prevention, and finally a dead reckoning technique for reducing the server load [2].

The game server is divided into a listener server and a dedicated server. The listener server runs as a process with the game client and can connect to a server hosted by another player without a separate server and play or invite other players. However, the disadvantage is that the server is also turned off when the game is turned off, so when the player ends the game, the host of the server is changed to another player. It is also a favorite way in LAN parties.

The dedicated server runs independently of the client process. It runs on a dedicated computer on a separate high-performance network, and players can enjoy pleasant gameplay by connecting to a dedicated server.

A dedicated server for large multi-user online games is a large-capacity game server operated by a specific company. You can run and maintain the server only by the developer who developed the game or by the developer's permission (so-called game publisher). In such a server, an unspecified number of players can enjoy the game together.

2 Related Works

2.1 Structure of Game Server

The first thing you have to decide before you develop any program is the programming language. In this paper, we propose C++ as a language for using IOCP. You can also build with C# or use .Net Frame Work with C#. .Net Frame Work is also implemented internally as IOCP, but the Garbage Collector in C# language has a big limitation. It is a big problem that there is a break in the game server where many packets are moving in real time. Using .Net Frame Work can increase productivity from the developer's point of view, but when it comes to commercialization, garbage-collector can cause server performance degradation. When using C++, it is possible to make low-level memory management compared to C#, so there is little overhead in writing and erasing the memory.

Multi-threaded use is inevitable to handle large amounts of packets received in real time from clients. The multi-threaded IOCP game structure [3] is generally in Fig. 2. Accept thread receives connections from clients through the thread. The connected client socket enters the asynchronous reception state after the information

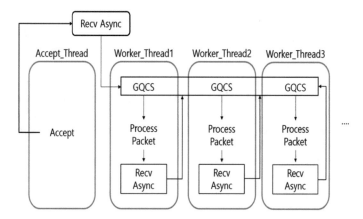

Fig. 2 Multi-threaded server architecture using IOCP

initialization operation registers it in the IOCP handle, and performs a full asynchronous I/O operation. When a packet is received for the socket, data is inserted
into the IO completion queue. Worker threads that are waiting are awakened by
the GQCS (GetQueudCompletionStatus) function. If there is data in the queue, the
GQCS function will wake up one of the worker threads randomly through the CPU
scheduler, take out the working distance, and process the packet for the packet. The
packet is processed and generated packets to be broadcast, and then the socket enters
the asynchronous receive state. Figure 2 is a minimum structure for communication,
and it is possible to add a thread to perform the function as needed by the developer.

In the case of a commercial game server, Fig. 3 has the form of a distributed
server structure [4]. If you do not use the distributed server architecture, the load
on the server increases, and as the number of users connected increases, you may
see a common "lag" phenomenon, such as connection failure or delayed input. The
client tries to connect through the login server, and the login server compares the
client's information through the DB server and then starts communication with the
main server. In the main server, user information is stored in the database in real time
through DB Server. In the MMO Game, you have to operate hundreds of thousands
of NPCs, and when you are in charge of the main server, the load on the server will
grow exponentially. Therefore, separation of AI Server from MMORPG is essential.
Figure 3 shows a functionally distributed server [5]. Such a server has an advantage
that the inconvenience caused by the service failure is reduced because the functions
other than the functions handled by the suspended server are operating normally.

Fig. 3 Commercial game
server architecture

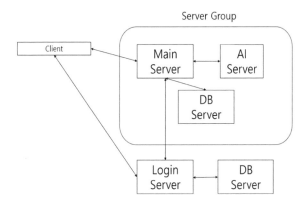

2.2 *Packet Design and Processing*

TCP conveys a byte stream that does not have a record boundary concept. TCP does
not have the concept of "packets" that users can see. This can be summarized as TCP
simply passing a stream of bytes. How to handle the flow of unbounded bytes as
much as the data sent by the user is explained. There are two ways to create a packet
on a server using the C++ language. The first is to use Array in Fig. 4 and the second
is to use structures and unions. There is no difference between the two methods. The
structured design has the advantages of good readability, but it has the disadvantage
of being dependent on C language and converting it when communicating between
different programming languages. Conversion work is cumbersome and error-prone
during the development process, so it is a prudent choice when communicating
between languages.

Fig. 4 Packet structure
using array

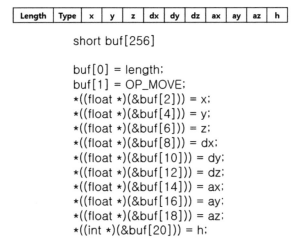

Packet Assembly Algorithm

```
void Packet_Assembly(id) {
int rest = io_size;
CHAR *buf  = overlapped->io_buf;
packet_size = overlapped->curr_packet_size;
old_received = overlapped->stored_packet_size;
while (0 != rest)
{
    if (0 == packet_size) packet_size = buf[0];
        required = packet_size-old_received;
    if (rest >= required) {
        memcpy(overlapped->packet_buf+old_received, buf, required);
        ProcessPacket (id,overlapped->packet_buf);
        packet_size= old_received = 0;
        buf += required;
        rest -= required;
    }
    else {
        memcpy(overlapped->packet_buf+old_received, buf, rest);
        old_received+= rest;
        rest = 0;
    }
}
overlapped -> curr_packet_size= packet_size;
overlapped -> stored_packet_size= old_received;
```

Fig. 5 Packet assembly algorithm

Most MMORPG games using IOCP use TCP. In TCP, the order is guaranteed [4]. However, as mentioned above, there is no boundary between packets because it is a stream of bytes. The first part of the packet specifies the length of the packet, and the second reason is that there is no boundary between the streams mentioned above. Figure 5 is an algorithm to assemble incoming packets into each packet. When the packet is assembled, the process is processed through the Process Packet function.

2.3 Optimization and Synchronization

The result of the data race is not a problem with a single thread [5], but it can be seen from Table 1 that an unintended value occurs in a multithreaded environment.

Table 1 Results for various numbers of threads

Number of threads	1	2	4	8	16
Results	100,000,000	52,232,532	31,536,756	17,579,488	7,133,668

Fig. 6 Sum operation with multithread causing data race

```
Packet Assembly Algorithm

sum
function Thread_func(thread_num)
{
    for i from 1 to 100000000/thread_num
        sum += 1
    end for
}
```

Figure 6 shows codes that can cause the data race. Table 1 shows the reason for the data race. This happens because one or more threads simultaneously read and write the shared variable Sum as shown in Fig. 6. In order to eliminate the data race phenomenon, it can be solved by using lock/unlock of the synchronization object in the critical region [6]. In [7], we compared the performance of synchronized objects when using two threads.

In this paper, we propose a visual processing technique and a dead reckoning [8] technique for MMO game server optimization. In the server, when the information is transmitted to all users when broadcasting, a considerable overhead occurs in the server. In the case of the MMO game server, a packet is required if the connection is N. As a solution for reducing the overhead, a method of broadcasting only object information in the vicinity can be solved. All game characters have the concept of sight. The game client also improves its performance by curling the frustum itself, and the server introduces the concept of view of the game character similarly and sends only the object information around the game character, which helps to improve the performance of the server and reduce the amount of network transmission.

Because of the large size of the MMO game, it is essential to efficiently search for nearby objects. A server that has nothing to do with the client has introduced the Sector concept that divides the map into its own [9], so that only the objects near the character can be efficiently searched by searching only the sectors near the player. If the sector is too large, many objects outside the scope of the search are searched, and because the threads share one sector in the search process, the parallelism is reduced.

If the sector is too small, there will be frequent sector changes when moving the character, and overhead will occur because the object list of the sector needs to be frequently updated. Dead reckoning is an obstacle to sending and receiving large amounts of data due to the existing network problems like bandwidth and delay. In this case, the client may appear to be disconnected from the screen.

In fact, it is a technique to make a delay happen but not to feel such a situation. Dead reckoning in the game is mainly used to update the user's location information. The most basic player movement method is to broadcast a mobile packet based on the user's key input, which causes a huge server load on a large-scale online game server due to its large load. Assuming a 10-s move in a 60-frame game, a total of 600 packets will be sent to the server. If there are ten players around, you have to broadcast 6600 packets including that user.

The dead reckoning technique broadcasts a packet to move through the user's first keystroke. There is no packet from the client until another keystroke arrives, and the server only sends the updated location periodically. This period can be determined by the status of the server or the nature of the game. The client can also update the location by comparing the position of the character and the position of the packet periodically coming from the server.

Dead reckoning in game development, but the words of a very narrow meaning referring to the situation that developed so difficult to progress in the game the error originated as screen inconsistencies among users due to the time delay now means the synchronization techniques and predictive algorithms to eliminate it [10].

The time delay is the time it takes a packet to travel from the client to the server and back to the client. Due to the time delay, users will see different screens, which disrupts the fairness of the game. Developers use dead reckoning to keep users from experiencing errors due to time delays and to maintain game fairness. In a large online game, a server processes hundreds or thousands of clients, and if the clients send frequent status update packets to the object, the server is overloaded with network loads. Therefore, if a large-scale online games, including the time delay, also resolved a letter to reduce the amount of packet [11].

There are three ways to update the state. The first is a method in which the client and the server periodically transmit state variables to each other. The second is a method of transmitting the state update packet only when the state changes to a predetermined value or more. The third is to send the update packet even when the status changes to more than a certain value for an instant response, by sending the update packet at regular intervals. Three methods are all playing the game according to the predictive algorithm based on the last received status information, if not received the update packet [8, 12].

The authors are concerned with a card game called Daihinmin (Extreme Needy), which is a multi-player imperfect information game [13]. The UEC Computer Daihinmin Championship is held at The University of Electro-Communications every year, to bring together competitive client programs that correspond to players of Daihinmin, and contest their strengths. Reference [13] extracts the behavior of client programs from actual competition records of the computer Daihinmin, and propose a method of building a system that determines the parameters of Daihinmin agencies by machine learning.

Comparing the wireless sensor network with other conventional network system is a tough job [14]. Therefore, since last decade the effort has been made to design and introduce a large number of a communication protocol for WSN with given concern on the performance parameter of energy efficiency and still the key requirements within WSN domain, that how to incrementally expands the energy minimization consuming techniques of sensor battery [14]. The other parameters include latency, fairness, throughput and delivery ratio. Reference [14] proposes a novel joint cooperative routing, medium access control (MAC) and physical layer protocol with traffic differentiation based QoS-aware for wireless sensor network (WSNs).

3 Implementation of a MMO Game Server

As following those in Fig. 3, this paper creates a simple MMO game server. For the use of libraries such as threads and synchronization objects, the language used was C++ 11, and the compiler used Microsoft Visual Studio 2015 [7]. The client also used the same compiler, and the graphics library used OpenGL [9]. The chess pieces are represented by moving chess pieces on the chessboard. The game server has eight threads excluding the main thread.

Packet processing and NPC logic processing, six worker threads for I/O, timer threads for continuously moving NPCs, and accesses threads for connecting users. The map was made at 500×500, and the size of flare and NPC was set to 1×1. We set the maximum number of users to 20,000 and put 5000 NPCs on the map.

When the server is first run, the NPCs are initialized. NPCs are randomly placed on the map, and the event to be acted on is inserted into the priority queue, which acts on the timer thread, with the id value, the time to act, and the action event. This priority queue determines the priority by comparing the time value for the NPC to perform the event.

Take the header part to the priority queue and let the WorkerThread process the NPC's logic through PQCS (PostQueueCompletionStatus). In the server, NPCs are set to move randomly every second. Table 2 shows the types of packets to be used.

Packets from the client to the server are packets indicating the direction through the key input. The server updates the direction in the worker thread according to the packet type through the ProcessPacket function, creates a packet in the SC_POS type, and broadcasts the location to the users in the field. The packet from the server to the client consists of SC_POS to inform the location, SC_PUT to let the player and NPC know when it is in sight, and SC_REMOVE to notify when the NPC or player is disappeared or disconnected. A visual processing technique was applied to the optimization technique used in the server. Dead reckoning is excluded due to

Table 2 Packet type	Server → client	Client → server
	SC_POS	CS_UP
	SC_PUT	CS_DOWN
	SC_REMOVE	CS_RIGHT
		CS_LEFT

Fig. 7 IOCP game server architecture

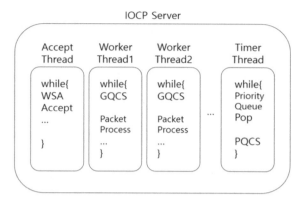

the nature of the chessboard client moving one space per keystroke. Figure 7 is a simplified version of the structure of the game server.

4 Performance Evaluation Using Stress Test

In this paper, we performed a performance test using a stress test program. Experimental items were an overload experiment in which dummy avatars were concentrated at a hot spot and a maximum dummy experiment in which dummy avatars were evenly scattered. The performance of the computer used in the experiment is as follows: Processor: AMD Ryzen 7 1700 Eight-Coire Processor 3.00 GHz, Memory 16.00 Gb, Operating System Windows 10 Pro 64 bit Operating System Graphics Card: Geforce Nvidia GTX1060 6 GB.

Stress test program was also produced through IOCP like the server created in Sect. 3. Draw Module that can grasp real-time location of connected dummy avatars and Network Module to communicate with the server in real time. Network module connected dummy avatar continuously for a specified maximum number of users and based on connected avatars. The key input packet is sent to the server every second.

4.1 Hotspot Simulation

All the connecting clients are designated as coordinate systems 10 and 10 and then connected. In Fig. 8a, the part where the central point is gathered is around 10, 10 which is set as the hot spot point. When the center of the stress test program window is set to 0, 0 and the 500 × 500 map is reduced to 1/400 scale, it is possible to check the position of the whole dummy avatars at a glance in real time.

After connecting about 900 dummies, the connection was slowed down due to the load in the test program. Over time, the number of connections slowed down

Fig. 8 Stress test

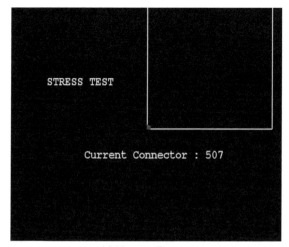

(a) Hotspot Test

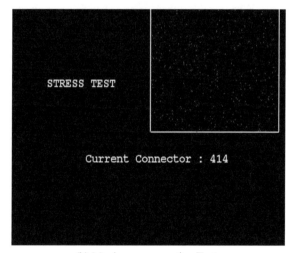

(b) Maximum connection Test

and the CPU utilization rises. Results of about ten tests are shown in Table 3. An average of 8721 dummy avatars received connections and did not receive connections from the server. After an average of about 8000 dummy avatars was connected, the CPU utilization was approaching 100%, and after the 8500 was exceeded, the time to connect increased by a factor of ten compared to just after the program was turned on.

Table 3 Hot spot test result using stress test

Test number	1	2	3	4	5
Connections	8743	8732	8674	8903	8873
Test number	6	7	8	9	10
Connections	8594	8602	8876	8562	8657

Table 4 Maximum concurrent connection test result using stress test

Test number	1	2	3	4	5
Connections	19,305	19,438	19,312	19,462	19,421
Test number	6	7	8	9	10
Connections	19,364	19,201	19,206	19,218	19,354

4.2 Simulation of Maximum Concurrent Connection

The connecting dummy avatars were placed randomly throughout the map. As shown in Fig. 8b, it is possible to confirm that the dummy avatars are dispersed in the map evenly. Due to the nature of broadcasting to N^2, the servers were overloaded quickly at the hotspot points, while the load of the servers was relatively late, so we could connect a larger amount of dummy avatars than hotspot tests.

This server could connect a much larger amount of dummy avatars than hotspot tests. On average, about 19,324 dummy avatars were connected, and we could see that the server overload was maximized as shown in Table 4.

Besides, we can confirm that there is little difference between the maximum and minimum samples compared to the hotspot test. This MMO game server was able to get a lot more connections than being expected while testing. It is expected that there is no complicated logic, and the movement of the NPC is very simple. Perhaps a game server that is built for service purposes is expected to connect far fewer avatars if simulated with the same specifications as the computer specifications tested in this paper.

4.3 Simulation of Maximum Concurrent Connection Except Visual Field Processing

In the maximum tangent test conducted in Sect. 4.2, the simulation was performed except for the visual processing technique. The results are shown in Table 5.

Except for the visual processing technique, approximately 6000 dummy avatars were not able to connect to the result in Sect. 4.2. It was part of the game server to see how important optimization was. In addition, various optimization techniques are expected to connect more dummy avatars.

Table 5 Maximum concurrent connection test result using stress test where visual field processing is excepted

Test number	1	2	3	4	5
Connections	12,817	12,938	13,031	12,962	13,011
Test number	6	7	8	9	10
Connections	13,014	13,031	12,986	12,998	13,054

5 Conclusion

In this paper, we propose the basic structure for MMO game server and suggest synchronization scheme and packet structure for performance enhancement. The techniques proposed are data race using synchronization objects, packet loss reduction using deadlock prevention, field processing, and dead reckoning. In the MMO game server where the number of broadcasting increases in proportion to the number of server accesses, the above-described view processing technique and dead reckoning technique are essential. Experimental results show that the performance improvement of the game server can be expected only by using the above two techniques. The experiments above were conducted with Local Host, so it is hard to trust absolutely on real commercial servers. Actual commercial servers will have more unpredictable parameters, and packet reception from clients will be more irregular than from experiments with stress tests.

Future research will focus on the use of multiple computers instead of a single computer to improve reliability in this paper.

Acknowledgements This work was supported by Institute for Information & communications Technology Promotion(IITP) grant funded by the Korea government(MSIP) (No. 2016-0-00204, Development of mobile GPU hardware for photo-realistic real time virtual reality).

References

1. Fall, K.R., Richard Stevens, W.: TCP/IP illustrated. In: The Protocols, vol. 1. Addison-Wesley (2011)
2. Loe, C.H., Seo, C.S., Wook, B.: Data priority-inheritance algorithm for deadlock prevention in distributed systems. In: Fall Conference of Korea Multimedia Society, pp. 106–111 (1998)
3. Choi, S., Park, H.: Study on the online game server architecture. In: Spring Conference of Korea Academia-Industrial Cooperation Society, pp. 534–538 (2006)
4. Jang, S.-M., Yo, J.-S.: An efficient MMORPG distributed game server. J. Korea Contents Assoc. **7**(1), 32–39 (2007)
5. Lee, N.-J., Gwak, H.-S.: The distributed server model for the evolutionary online RPG. J. Korea Game Soc. **2**(1), 36–41 (2002)
6. Savage, S., Burrows, M., Nelson, G., Sobalvarro, P., Anderson, T.: A dynamic data race detector for multithreaded programs. ACM transactions on computer systems **15**(5), 391–411 (1997)

7. Aggarwal, S., Banavar, H., Khandewal, A., Mukherjee, S., Rangrajan, S.: Accuracy in dead-reckoning based distributed multi-player games, NetGames '04. In: Proceedings of 3rd ACM SIGCOMM workshop on Network and system support for games, pp. 61–165 (2004)
8. Shim, K.-H., Kim, J.-S.: A study on performance analysis and improvement of dead-reckoning algorithm in networked virtual environment. Fall Conf. Korean Inst. Inf. Sci. Eng. **28**(2), 112–114 (2001)
9. Seok-Jong, Yu.: Game server and spatial partitioning for MMORPG. Comm. Korean Inst. Inf. Sci. Eng. **23**(6), 29–35 (2005)
10. Kim, S.-R., Yun, N.-K., Koo, Y.-W.: Design and implementation of dead reckoning algorithm for network game. Korea Inf. Process. Soc. **7**(8) (2005)
11. Kim, K.-C.: Online Game Server, EGO, 141–158 (2012)
12. Lengyel, E.: Believable Dead Reckoning for Networked Games, Game Engine Gems2, A K Peters, 307–327 (2011)
13. Wakatsuki, M., Fujimura, M., Nishino, T.: A decision making method based on society of mind theory in multi-player imperfect information games. Int. J. Softw. Innov. (IJSI) **4**(2), 58–70 (2016)
14. Haqbeen, J.A., et al.: Design of joint cooperative routing, MAC and physical layer with QoS-aware traffic-based scheduling for wireless sensor networks. Int. J. Netw. Distrib. Comput. **5**(3), 164–175 (2017)
15. Moon, S.-W., Cho, H.-J.: A study on synchronization distribution of server message in online games. J. Korea Game Soc. **9**(2), 105–113 (2009)

A Study on the Characteristics of Electroencephalography (EEG) by Listening Location of OLED Flat TV Speaker

Hyungwoo Park, Sangbum Park and Myung-Jin Bae

Abstract Along with the development in OLED TV panel manufacturing technology, the thickness of the panel is thinned, and the durability is also improved. It is also possible to generate sound by directly driving the panel. In this study, we have analyzed the acoustic characteristics of a directly driving the OLED panel speakers, using electroencephalography (EEG), in addition study the advantages of direct driving sound. For the experiment, we found that it is important that the two focal points coincide with each other through the EEG analysis when the focus of the screen and the focus of the sound are changed by adjusting the sound generated by the position in the OLED flat panel TV speaker. OLED TV with direct drive speakers has a basic effect of matching the positions of the screen and sound. This is because the direction of the sound and the direction of the screen coincide with the viewers of the TV, so that the stress is less and the viewing is comfortable compared to the conventional TV. The acoustic characteristics of proposed OLED speakers are high in resolution. And the sound quality is better than ordinary TVs, because of direct transferred sound instead of listening to reflections or difference sound sources compared with.

Keywords Direct drive speaker · Exciter speaker of OLED TV · Stereo excite speaker · Electroencephalograph (EEG) test

1 Introduction

People get a lot of information around them through visual, auditory, olfactory, tactile and, taste sensations. It also produces information, recognizes information and acquires information from these five senses. Especially, audiovisual is a very impor-

H. Park (✉) · S. Park · M.-J. Bae
School of Information and Technology, Soongsil University, Seoul, Republic of Korea
e-mail: pphw@ssu.ac.kr

S. Park
e-mail: sbpark8510@naver.com

M.-J. Bae
e-mail: mjbae@ssu.ac.kr

© Springer International Publishing AG, part of Springer Nature 2019
R. Lee (ed.), *Computational Science/Intelligence & Applied Informatics*, Studies in Computational Intelligence 787, https://doi.org/10.1007/978-3-319-96806-3_7

tant information transmission factor and has been developing various technologies and methods for a long time [1]. And people wanted to convey information by transferring sound and image far away, and they studied the technology and device to convert information form through electric/electronic devices, photography, recording, amplification device, etc. and stored real image and sound in the very same form, Transmitted, and reproduced [2]. Among the technologies, TV is a complex of advanced technology that transmits and reproduces video and sound, and it becomes an inseparable entity in people's lives [3].

With the development of IT technology, black and white video, mono sound TV, color video, stereo sound source, High Definition (HD) and Ultra High Definition (UHD) have been developing remarkably. The development center of this TV has mainly been in the direction of design and image, and it has been developed by changing the way of making a light and forming an image, such as CRT, LCD, PDP, LED and OLED [4]. And it has been developed to realize 3D image on 2D screen and to experience virtual reality and augmented reality [5]. However, in the design center, it is difficult to reduce the space for producing the sound, or to hide the speaker to realize high quality sound. Therefore, this study proposes a crystal sound OLED that can reproduce high quality sound by matching sound position and sound direction on the screen, and proves the effect. In this study, we have the experiment of EEG test when watching TVs, with ordinary OLED TV and proposed OLED TV at the same time. We analyzed the change of EEG, during watch both TVs, and the influence of the coincidence of screen and sound on the increase of absorbedness was judged [6].

In the early CRT TV, it was surprising to the world to transmit monochrome images and monaural sound through electromagnetic waves, and to reproduce videos transmitted from a remote place. With the advancement of technology, analog color video and stereo sound can be sent and received in the same electromagnetic band, and technology is developing and HD, UHD, DMB, and HD streaming are growing remarkably [2]. In this development process, the pixels constituting the screen are changed from cathode-ray tube (CRT), a liquid crystal (LCD), to a light itself (LED, Light Emitting Diode), and a pixel size is reduced [3, 4]. Along with the development of the pixel itself, the form of the screen composed of the pixels is also changed. In the initial TV model, both the CRT and the speakers constituting the screen were directed to the front, so that at least the direction of the sound and the image coincided. Since then, as the technology that makes up the screen has been developed, the bezel that fixes the screen gradually narrows and disappears, and the information display technology improves, and its thickness becomes thinner from 1 to 2 cm–2 to 3 mm. In the OLED information display device, an ultra-thin display device in which an OLED layer constituting an actual screen is not more than 1 mm is developed and put on the market. In addition to this, the technology has been developed to improve the information display so that 3D can be felt on the 2D screen, and to fold or curved the display to make the natural screen configuration [5].

However, the development of the above-mentioned display has caused the abstraction, disconnection, and suppression of TV sound. The speaker, which plays the sound, is hidden behind, on the side, and on the bottom of the screen, emphasizing the display and appearance design of the display. This breaks the image and sound of early CRT TVs and prevents the development of high-quality sound technology. Depending on the principle of sound generation, the sound is knocked or frictioned, and the resonance of the surroundings is gained to be amplified loudly [7]. Resonance, in particular, receives a characteristic of the space in which sound is generated and sounded [8]. As the vacant space of the display device, that is, the space in which the speaker can resonate, is reduced, the probability of reproducing high quality sound is reduced accordingly [9]. This trend has led to the development of high-quality sound and high-quality images. Especially, in order to reinforce the low-volume real sound system built-in the display device, the policy of the home appliance companies which separately make and distribute the sound bar and the home theater system separately cannot mix the two technologies.

Traditionally, the way to produce sound is to tap, bounce, rub, and blow in air. These methods are difficult to reproduce after storing and storing sound, and distant delivery is also a disadvantage [10]. To overcome this, people began to convert the sound into a different form that could be stored or stored there, with voices using letters and music using music. However, the development of IT technology allows the generated sound to be restored and reconstructed. It is possible to convert the sound into an electrical signal with a microphone, store the changed electric signal as digital or analog, then reproduce it through the amplifier and speaker again. The microphone and the speaker are opposite to each other, and the electric signal stored or transmitted by the speaker is converted into acoustic information from the kinetic energy and propagated through the air. Therefore, the speaker consists of a magnet, an electromagnet, a diaphragm, a structure for holding the whole, and a connector for smoothly moving them, in order to spread the electric signal to the air by the vibration of the vibrator. Conventional loudspeakers use thin, rounded paper as the diaphragm, which is called a cone. And it can be a good speaker if you attach it to the cone and coil so that you can move smoothly. However, the speaker used in this study is greatly changed by using the OLED panel as the diaphragm [11].

In this study, we have the experiment of EEG test when watching CSO, with a different sound position which is divided 9 zone on the screen of plat panel display. We analyzed the change of EEG, during watching 9 zoned the screen and different sound position, and the influence of the coincidence of screen and sound on the increase of absorbedness was judged.

2 Related Works and Background Principal

2.1 Speakers and Sound

Since the development of a speaker as a converter that converts an electric signal into an acoustic signal by converting an object into an acoustic signal, the development of a magnetic material such as a permanent magnet after improvement of high-temperature adhesive and related materials has been developed in the early 1930s,

Fig. 1 Ordinary dynamic
speaker [2]

respectively [6]. The most common type of driver, commonly called a dynamic loudspeaker, uses a lightweight diaphragm, or cone, connected to a rigid basket, or frame, via a flexible suspension, commonly called a spider, that constrains a voice coil to move axially through a cylindrical magnetic gap. When an electrical signal is applied to the voice coil, a magnetic field is created by the electric current in the voice coil, making it a variable electromagnet. The coil and the driver's magnetic system interact, generating a mechanical force that causes the coil (and thus, the attached cone) to move back and forth, accelerating and reproducing sound under the control of the applied electrical signal coming from the amplifier. The following is a description of the individual components of this type of loudspeaker. Edge: It is important for the unit to shake the cone paper. That used paper before, but nowadays use the edge of rubber and synthetic resin. Gasket: It is between the edge and the frame and fixes the edge so that it does not get out of the unit. Diaphragm (cone, cone paper): The cone paper is shaking and the air is vibrated. Plates or synthetic materials are used to prevent moisture from being vulnerable. Frame: Fix the unit to the skeleton of the speaker unit. Cap: Blocks foreign objects entering the speaker. Damper: It is located between the diaphragm and voice coil and controls vibration of the diaphragm. Voice coil: Receives a voice signal and transmits it to the diaphragm. The higher the density, the better the quality. Magnet: Helps the voice coil moves up and down. It affects negative pressure and vibration. Plate: Allows magnetic force to pass through [9] (Fig. 1).

Producing a sound from a speaker is made by moving the electromagnet in the magnetic field and trembling the cone as much as the movement. With the development of speaker systems, dynamic speakers can reproduce the original sound. However, dynamic loudspeakers have a disadvantage in that they require a large volume for reproduction, because they produce various frequencies using the vibration of the cone and the resonance of the enclosure. Various types of diaphragm shapes have been proposed and developed to compensate for these drawbacks. In addition to the voice coil, a form has been developed to produce vibrations in different ways. And there have been many attempts to reduce the size of the speaker systems, or alternatively to make them less obvious. One such attempt was the development of "exciter" transducer coils mounted flat panels to act as sound sources, most accu-

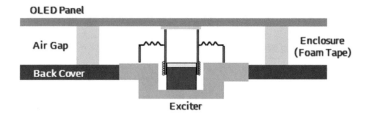

Fig. 2 A model of flat panel speaker [7]

rately called exciter/panel drivers. The thickness can be made very thin compared to dynamic speakers. However, it is difficult to make good sound due to the limitation of the appearance, and there is a disadvantage that the quality of the sound changes greatly with the shape and the material of the diaphragm. In addition, the distortion is large and the control is difficult because of small resonance at enclosures. Particularly, a more flexible and lightweight material is required compared to a conventional diaphragm, Cone-paper, because the exciter-generated vibration must be transmitted through to the diaphragm. In order to control the vibration of the plate by a single exciter, there is a limitation in making good sound with the vibration mode control of the plate and the reflection wave according to the impedance difference at the edge of the diaphragm. Figure 2 can be modeled as a 1 degree-of-freedom vibration system in which exciter, back cover, and flat plate are connected in harmony [7, 9, 11].

2.2 Electroencephalography (EEG) Test

"Electroencephalography (EEG) is an electro-physiological monitoring method to record electrical activity of the brain" [9: p. 94]. "It is typically non-invasive, with the electrodes placed along the scalp, although invasive electrodes are sometimes used in specific applications" [9: p. 95]. The EEG measures voltage fluctuations due to ion currents in the brain's neurons [9]. In the clinical context, EEG refers to recording spontaneous electrical activity of the brain over a period of time as recorded on several electrodes placed on the scalp [9]. Diagnostic applications generally focus on the type of neural vibration called the 'brain wave' seen in the EEG's spectral content, the EEG signal [9, 10].

EEG is typically analyzed to distinguish a total of five to Delta wave, Theta wave, Alpha wave, Beta wave, Gamma wave. And Table 1 shows the characteristics of the state of the frequency and the brain for the five kinds of EEG. Generally, brain waves is obtained in the form of continuous waves. After the frequency analysis with the waveform analyzes so, the state as shown in Table 1. EEG is not represented by 100% in entirely the form of any particular period. A wavelength depending on the situation and also represents a configuration in a ratio. In another analysis, it is determined by the absolute energy is above a certain amount or the like may use the results [12].

Table 1 Types and features of the EEG [12]

Indicator	Frequency definition (Hz)	State
Delta wave	0.1–3	Deep sleep
Theta wave	4–7	Sleep
Alpha wave	8–12	Awake
Beta wave	13–30	Tension, excitement, stress
Gamma wave	30–50	Anxiety, nervous, stress

The EEG is most often used to diagnose epilepsy, which causes abnormalities in EEG readings. It is also used to diagnose sleep disorders, coma, encephalopathies, and brain death. EEG used to be a first-line method of diagnosis for tumors, stroke and other focal brain disorders, EEG continues to be a valuable tool for research and diagnosis, especially when millisecond-range temporal resolution is required [13].

3 Proposed Direct Drive Plat Panel Speaker

Unlike conventional flat-panel loudspeakers, in this study, we use directly oscillates the display layer of the OLED that used as the diaphragm of the speaker. The development of display devices, OLEDs can now use a single glass layer to construct pixels of the panel. A flat panel display device can be roughly classified into an LCD and an LED. In the case of an LCD panel, a number of optical sheet layers, a layer constituting a pixel by a liquid crystal and a liquid crystal cannot emit light by themselves. Is composed of many layers of film, liquid crystal, backlight, etc. up to the layer and the layer, causing reflection [14]. This structure is disadvantageous to transmit vibration even if it is made thin, and even if the actual display panel is vibrated, vibration damping occurs into the layer, so that sound is not appropriately generated. Furthermore, liquid crystals contain disadvantages such as an easy color change in physical vibration, shortened life span. On the other hand, LEDs constitute pixels by themselves and generate light itself, and since there is no backlight layer and pixels are formed by using the three primary colors of light, an optical sheet corresponding to the filter on the front panel can be omitted [15]. Therefore, it is easy to vibrate the panel itself, and the attenuation of transmission can be reduced, and the sound can be efficiently reproduced. The structure of the LCD panel and the LED panel and the transmission of the vibration are shown in Fig. 3. As shown in figure, the LCD is composed of various layers, and the OLED is composed of a simple layer and the vibrations generated from the exciter are displayed on the screen [16].

As shown in Fig. 3, even if the OLED panel vibrates directly to produce sound, the sound quality varies greatly depending on the display shake characteristics and vibration mode characteristics. As we have seen for flat speakers in Chapter "Python Deserialization Denial of Services Attacks and Their Mitigations", the sound is

Fig. 3 The plat panel display consists with the exciter

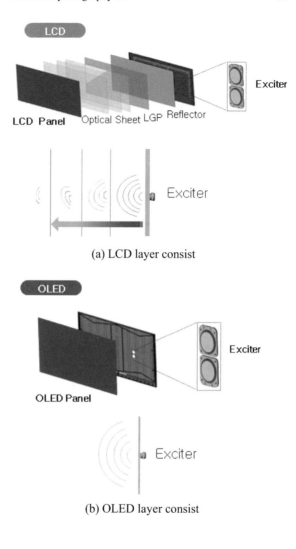

(a) LCD layer consist

(b) OLED layer consist

different depending on the shape and characteristics of the diaphragm. In order to compensate for the disadvantages of the conventional flat speaker, it is necessary to improve the sound quality by transmitting the vibration to the back plate which fixes the display [17]. In order to realize the sound of the left and right stereo channels moving in conjunction with a single plate, the technique of preventing the vibrations generated on the left and right sides from affecting each other was also applied. Using this method, one glass layer is used as a diaphragm, but a flat speaker in which stereo sound is reproduced can be constituted. These studies have been carried out in the preceding researches, and the study on the improvement of the sound quality according to the location of the exciter has also been carried out. In this study, we evaluate not only the optimum position of the exciter, but also the effect of

Fig. 4 The virtual planes
occupying an even area of
panel

1	2	3
4	5	6
7	8	9

encodings through adjustment of the acoustic impedance with the back plate, and the transmission of information according to the composition of the screen and the composition of the sound. For this experiment, the screen is divided into nine small virtual planes occupying an even area as shown in Fig. 4, and the perceived change according to the sound and visual position generated on the surface is measured.

4 Experiment and Result

We measured and compared the sound quality of different sound positions in the proposed OLED TV set which is a direct-drive exciter speaker. For this experiment, we generated sounds at nine virtual positions, set the same weather announcement on one screen, and measured sound and EEG by location. To do this, I used a 10 W exciter speaker and a typical flat panel TV amp. Basically, it was confirmed that the sound toward the front same as the direction of the screen is clearly reproduced unlike the conventional general flat panel TV. The quality of the sound reproduced at each position was confirmed to be very uniform according to the shape of the well-tuned panel. Experiments were performed to see the same video 9 times and the sound position was changed from 1 to 9 in Fig. 5. Figure 5 shows the change of EEG according to the time. As shown in the figure, the measurement results of the upper and lower brain waves are very different. The upper part is the case where the sound position is different from the mouth position of the person on the screen, and the lower figure is the brain wave when the mouth position and the sound position of the person are matched. When we use the change of EEG composed by time, we can confirm that people do not concentrate on the screen when the sound of the upper part and the focus of the screen are different.

The concentration of energy in the vicinity of the low frequency is the result of the EEG obtained by not concentrating on the image and sound appearing on the screen while listening. When the δ and θ waves corresponding to the slow-wave are strong energies, it corresponds to a state where the human brain is resting, sleeping, or no concentrating, while watching different position of sound and screen. However, when watching matched sound and screened TV, it is necessary to watch EEG more than EEG (low frequency) rather than EEG (high frequency EEG). In the case of

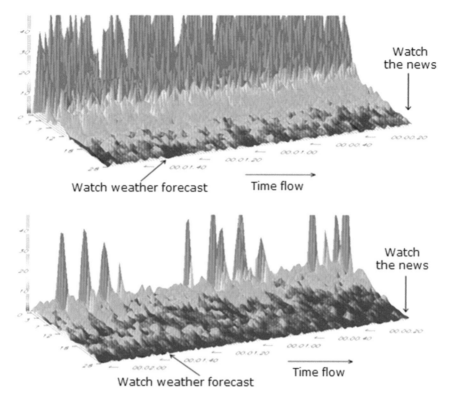

Fig. 5 Experiment result of EEG, compare with watching TVs

the different position, the low frequency corresponding to the west of the total EEG energies occupies a lot of parts, which means that the movement of the EEG becomes very slow and rest and dormant state when watching TV. In other cases of sound and picture is matched, is more focused than the mismatched situation. And the sound and the screen match well and the burden on the brain is less.

5 Conclusion

Flat panel TVs are getting thinner due to advances in display technology and semi-conductor technology. And self-illuminating OLED technology makes the picture better, and the sound is enhanced using the panel. Also, in the previous research, we proposed OLED display with matching screen and direction through CSO. In the previous research and this study, it is a good way to make the sound by directly vibrate the screen of the flat panel display to match the position of the screen with the position of the sound. In this study, it is shown how effective it is. In this study, we measured the change of sound quality and the change of EEG when the position of sound was changed by playing the same screen. As in the previous study, it was confirmed that EEG when the positions of the screen and sound coincide steadily acquires information. As a result, the proposed TV, which can match the position of video and sound, has excellent immersion feeling and can feel the accurate sound field, showed excellent results. In the future, we will study the vibration characteristics and the acoustic characteristics to improve the sound quality and make the three-dimensional sound field feel.

References

1. Kim, J.T., Kim, J.H., Kim, J.O., Min, J.K.: Acoustic characteristics of a loudspeaker obtained by vibroacoustic analysis. Trans. Korean Soc. Mech. Eng. A **21**(10), 1742–1756 (1997)
2. Choi, H.W., Kim, Y.J., Park, Y.W.: Acoustic analysis and vibration modeling for design of flat vibration speaker. In: Proceedings of Korean Society for Precision Engineering, 2008 Fall Conference, pp. 545–546 (2008)
3. Park, H.W., Bae, M.J.: A study on the improvement of sound quality according to the location of OLED flat plate speaker. Asia Pac. J. Multimed. Serv. Converg. Art Humanit. Sociol. **7**(12), 775–783 (2017)
4. Park, H.W., Bae, M.J.: A study on the watching attention according to the sound position of flat panel TV. Asia Pac. J. Multimed. Serv. Converg. Art, Humanit. Sociol. **7**(7), 839–846 (2017)
5. Choi, D.J., Park, Y.W., Park, H.J.: Research of relation between sound pressure and magnetostrictive speaker for 2ch flat display. In: Proceedings of Korean Society for Precision Engineering, 2010 Fall Conference, pp. 603–604 (2010)
6. Jung Beon, H.: Trends in speaker manufacturing technology at domestic and abroad. Mag. IEEE Korea **13**(6), 513–520 (1986)
7. Nam, S.H., Choi, D.J., Chae, J.S.: Objective audio quality assessment. Proc. IEEE Korea, 108–111 (1995)
8. Kim, J.H., Kim, J.W.: Thin speakers and transparent speakers. CERAMIST **17**(2), 60–66 (2014)
9. Oh, S.J.: Theory and Design of Loudspeaker. SuckHakDnag press (2011)
10. Cheon, E.: Neurofeedback treatment in adult psychiatric patient: focusing on depressive and anxiety disorder. J. Korean Soc. Biol. Ther. Psychiatry **19**(2), 85–92 (2013)
11. Lee, H.J., Shin, D.I., Shin, D.K.: The classification algorithm of users' emotion using brainwave. J. Korean Inst. Commun. Sci. **39**(2), 122–129 (2014)
12. Lee, S.: Basic Properties of Sound and Application. Chung-Moon-Gak Publisher (2004)
13. Park, U., Lee, S.B., Lee, S.H.: General Theory of Sound Technology. Cha Song Publisher (2009)
14. Bae, M.J., Lee, S.H. (eds.): Digital Speech Analysis. Dong Yeong Publisher (1998)
15. Kown, O.G. (ed.) Design of Multi-Channel Speaker System Using Digital Audio Technology. Hankukhaksuljungbo (2008)
16. Haykin, S. (ed.) Array Signal Processing, 493 pp. Prentice-Hall, Inc., Englewood Cliffs, NJ (1985), For individual items see A85-43961–A85-43963 (1985)
17. Kim, Y.H., Nam, K.U.: Lecture Notes on Acoustics. Chungmungak pass (2005)

A Study on the Design of Efficient Private Blockchain

Ki Ho Kwak, Jun Taek Kong, Sung In Cho, Huy Tung Phuong
and Gwang Yong Gim

Abstract Blockchain is the key innovative new technology of the future for which interest is being increased domestically and overseas as an icon of innovation to change the existing business process. The Blockchain has started as a public Blockchain first, allowing anybody to participate in the Blockchain network to view all history, and anybody to verify transaction history. However, as proved by bitcoin, the open-type distributed ledger has several technical and positive problems. There exist several disadvantages where much resource should be incorporated to maintain and manage the network with unspecified many participating, internal information is disclosed, processing speed is slow, and anonymity of transactions should be guaranteed, etc. The private Blockchain technology that overcomes such limitations and configures a distributed network with only certified participants as the subject is being magnified. In the present article, issues of a reliable private Blockchain will be derived and design measures be proposed to make a contribution so as to enable safe utilization in the environment of P2P distributed network.

Keywords Blockchain · Information protection · Architecture · Security
Certification

K. H. Kwak · J. T. Kong · S. I. Cho · H. T. Phuong · G. Y. Gim (✉)
Department of Business Administration, Soongsil University, Seoul, South Korea
e-mail: gygim@ssu.ac.kr

K. H. Kwak
e-mail: khkwak1004@gmail.com

J. T. Kong
e-mail: jtkong31@gmail.com

H. T. Phuong
e-mail: tungph@soongsil.ac.krcom

S. I. Cho
Department of IT Policy and Management, Soongsil University, Seoul, South Korea
e-mail: cho5392@naver.com

© Springer International Publishing AG, part of Springer Nature 2019   93
R. Lee (ed.), *Computational Science/Intelligence & Applied Informatics*, Studies
in Computational Intelligence 787, https://doi.org/10.1007/978-3-319-96806-3_8

1   Introduction

Blockchain known as the basic technology for bitcoin is receiving attention as the innovative security technology of a connected society.

While most of the existing centralized systems maintain security using powerful access control through diversified equipment and software, Blockchain maintains security by opening this. In addition, Blockchain technology does not require the central server. Since it is operated in a P2P distributed network, costs for setting, maintenance and management are greatly reduced [1].

It is expected to innovate application programs and promote the second internet revolution with a potential to redefine digital economy [2].

Although the Blockchain technology in the 4th industrial revolution era has an aspect of being widely known and partly swollen due to bitcoin cryptocurrency, domestic and overseas interest is being increased much as the icon of innovation that will change the existing business process with the future key new technology of the future due to the expectation that it will provide information sharing method with a new paradigm.

IDC recently forecast, stating "Blockchain ledger and interconnection will be developed consistently for 36 months up to 2021, and at least 25% of global 2000 enterprises will use the Blockchain as a foundation for large-scale digital reliability" [3].

Blockchains are a remarkably transparent and decentralized way of recording lists of transactions. Their best-known use is for digital currencies such as Bitcoin, which announced Blockchain technology to the world with a headline-grabbing 1000% increase in value in the course of a single month in 2013. This bubble quickly bursts, but steady growth since 2015 meaning Bitcoins are currently valued higher than ever before.

A series of startup enterprises together with IBM, Microsoft are already developing a considerable scale of business in relation to the construction of Blockchain platform, and competitively seeking the methods for saving time and administration costs through the Blockchain technology.

The Blockchain is largely divided into public Blockchain and private Blockchain. While public Blockchain means the Blockchain where anybody can participate in the network as with bitcoin or Ethereum, private Blockchain means the Blockchain used independently by an institution.

The Blockchain has started first as allowed to participate in the Blockchain network and view all history and can verify transaction history. However, as proven by the bitcoin, the open-type distributed ledger carries several technical and positive problems. There exist several disadvantages where much resource should be incorporated to maintain the network participated by unspecified many, internal information related to transfers disclosed, processing speed is slow, and anonymity of transactions should be guaranteed, etc. Thus, the private Blockchain technology is being magnified where such limitations are overcome and a distributed network is configured only with the certified participants as the subject [4].

The private Blockchain also called as permit-type distributed ledger maximizes security and safety while overcoming the limitations of the existing public Blockchain.

Domestic financial institutions and major large conglomerates are implementing technology verification for the private Blockchain, through which they are actively reviewing incorporation thereof as an innovative measure for values of the enterprise.

However, the private Blockchain is not free of any disadvantages, but requires studies for sustained improvement by finding supplementation points.

2 Theoretical Backgrounds

2.1 Definition of Blockchain

Blockchain is a compound word for "block" meaning connection and "chain". Blockchain has been proposed in 2008 by Satoshi Nakamoto, and first introduced to the world with the emergence of bitcoin. Each block containing data is connected to the previous block through a chain, with the Blockchain, displaying a structure with such data linked. Namely, the Blockchain may be considered as a form of distributed database where the structure for data storage and processing is configured by block connection. However, the Blockchain may be defined in diversified ways depending on viewing methods. In the present study, the Blockchain is of a structure where the blocks (data) with validity inspected are linked, and defined as the technology for distributed the database where data is jointly recorded managed at P2P network participant (node) [5].

2.2 Features of Blockchain

The greatest feature of the Blockchain is being a network-based distributed system and a structure that is operated without a central management institution playing the role of an intermediary for transactions (Fig. 1).

The Fig. 1 allows a very intuitive description of characteristics of the Blockchain, where most systems are currently composed of centralized systems as shown in the left-side figure, expressing that all services are realized through the managing entity having data, i.e. the central control system. On the other hand, the Blockchain-based system on the right side has a structure where data is directly transmitted and received among individual nodes participating in the network without the central control system as the managing entity, and information related to this is shared.

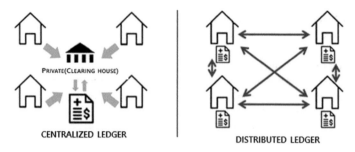

Fig. 1 Features of Blockchain system

2.3 Structure of Blockchain

Data structure of the Blockchain is a structure containing transaction information in the block with the relevant block being connected to the previous and the following blocks. A block is composed of block size (4 byte), header having Meta information (80 byte), and information on the transaction recorded in the block. Whenever a transaction occurs, a block of the transaction record is created and linked. The transaction is transmitted to all participants, and the validity of the transaction is approved. The approval method is a proof of work. The block that has been validated through the proof of work refers to the previous block and linked to a new block. As the chain becomes longer, the reliability increases since it is difficult to forge or falsify if there is a large amount of block connected (Table 1).

Hash value of the previous block is included in the block header, and arrangement of the hash values connecting the current block and the previous block, i.e. the parent block constitute a chain connected to the first block. 32 byte value generated by encoding of the block header through SHA256 encoding hash algorithm is called the block hash, with this value identifying the relevant block, and all participant nodes allow hash value to be obtained by generating the block header [6] (Table 2).

In the connecting structure of the block, the block has an identifiable hash value for header information of the previous block, while the block has a structure connected by referring to the value. The longer the chain, the more increased is the reliability if many blocks are connected, since it is a structure that makes forgery difficult [6].

Table 1 Block structure

Size	Field	Description
4 byte	Block size	Size of block
80 byte	Block header	Meta information recorded in block
1–9 byte (varInt)	Block transaction no	No of block transactions
Variable	Transaction record	Transaction recorded in block

Data: Antonopoulos (2015)

Table 2 Structure of block header

Size	Field	Description
4 byte	Version	Version no for tracking of software/protocol upgrade
32 byte	Hash of previous block	Value allowing reference to the hash data of previous block/parent block that the chain has
32 byte	Merkle root	Root of Merkle tree included in transaction information that the relevant block has
4 byte	Time stamp	Time for block generation
4 byte	Difficulty target	Difficult target of difficulty for algorithm that the operation proof method has
4 byte	Nonce	Counter applied to the algorithm that the operation proof method has

Data: Antonopoulos (2015)

2.4 Principle of Blockchain

Blockchain is the database structure where the blocks connected to the chain exchange information whenever a new transaction occurs in the existing block. It is the method where transactions are approved for all participants connected to the network and the approving method undergoes the process called operation proof.

Operation proof is a very important element in the Blockchain, through which security and reliability in Blockchain transactions are increased (Fig. 2).

In the structure of Blockchain database, a new block including the new transaction history is formed at a predetermined time and connected to the existing block. This transaction record is not saved and managed in the central server but shared by all

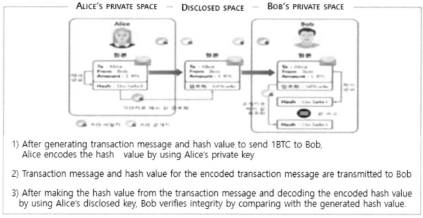

1) After generating transaction message and hash value to send 1BTC to Bob, Alice encodes the hash value by using Alice's private key

2) Transaction message and hash value for the encoded transaction message are transmitted to Bob

3) After making the hash value from the transaction message and decoding the encoded hash value by using Alice's disclosed key, Bob verifies integrity by comparing with the generated hash value.

Data: Financial Security Institute (2016)

Fig. 2 The principle of block transaction

participants. Whenever a transaction occurs, a transaction record block is generated and connected. The transaction is transmitted to all participants and validity of the transaction history is approved. According to this approval method, the block with validity verified through operation proof refers to the previous block and is linked to the new block. When the chain becomes long, for connection of a large number of blocks, forgery is difficult so that reliability is increased [7].

Blockchain is safe from hacking or outside invasion due to the distributed structure so as to provide a high reliability. Its data structure is transparent, making forgery almost impossible. Since operation occurs with the data being shared at all nodes of network, there is no spot for a single failure, while the system is stable and available. Blockchain can be used in diversified areas, being characterized by diversity. Since the use of Blockchain allows reduction of costs for construction, maintenance and repair through the distributed system, the Blockchain is economical [8].

3 Precedent Studies

3.1 Types and Features of Blockchain

Blockchain is largely classified into Public Blockchain, Private Blockchain and Consortium Blockchain [6]. Public Blockchain may be considered as a general Blockchain technology with openness and dispensability in which anybody can participate as with bitcoin. On the other hand, private Blockchain is a privatized Blockchain that one entity controls with authority. It is a system for authorized users, and has limitations in the operation nodes. Consortium Blockchain allows transaction when there is consent/agreement on the nodes operated by n entities although it is controlled by the nodes already agreed upon [6] (Fig. 3 and Table 3).

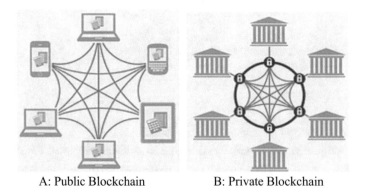

A: Public Blockchain   B: Private Blockchain

Fig. 3 Public and private Blockchain

Table 3 Types of Blockchain

Classification	Concept and feature	Remarks
Public Blockchain	First case of Blockchain utilization Transaction ledger disclosed all involved parties of transaction through internet Everyone can participate in notarization by using computer power Problems if difficulty in transaction expansion and speed delay	Network effect Stable ecosystem Risks for 51% attack or dual remittance exist
Private Blockchain	Private-type Blockchain concept Structure where on management entity can exercise management and authority Emergence of private service	Supervision according to the risks of system change and securing of safety required Global branch secured through one entity
Consortium Blockchain	Semi-centralized Blockchain Controlled by selected nodes with 1 ea. Of node operated by n entities easy expansion of network and fast transaction speed	Business consent/agreement among participating entities System stability secured No problems of 51% attack or due remittance exist

Data: Oh Seyong (2017)

Before application of Blockchain, the technology should be utilized by considering the characteristics of each Blockchain as well as the characteristics of the contents to be serviced by the enterprise. Also, all services cannot be naturally considered as the Blockchain. If concentration of all data in the center for service is efficient, the Blockchain needs not be considered. The current centralized service can be more advantageous. And service needs to be provided by considering the subject to be used and selecting among the diversified methods of Blockchain [6].

3.2 Problems with Private Blockchain

Blockchain technology has a potential capable of innovatively changing reliable models and business processes in various industry areas. However, this technology is still at an initial stage, and distributed ledger technology used in the Blockchain technology is not receiving appropriate monitoring or checkup.

Blockchain is a system that grants autonomy while abandoning efficiency. If a new block is to be added to the Blockchain, encoded inspection procedure for all blocks is required. For such reason, application of prompt transactions to an essential business area is not efficient.

The Blockchain technology is still a vague concept in which the Blockchain handles security using encryption techniques, but it is different from encryption. For the default, the contents, transactions and records of the Blockchain itself are not encrypted. The reason why bits are secured with the encryption technique is to immediately determine the hash and connection after the user harpoons the transaction and attempts to modify the hash. This is because the hash does not match.

One of the key features of the Blockchain that financial services and insurers seek is to ensure transactional security and reduce risk. That is because when a record is changed, the person who checks the chain can immediately know it. But it might also give up information. Because the contents based on Blockchain is pure text, although it may be a bit difficult but it could be easily encrypted. Since the Blockchain is literally in the form of 'chain', block insertion should be serialized. Hence, the update speed is slower than with the traditional database subjected to parallel update.

Recently, an accident occurred in the Blockchain that a user froze Ether as Ethereum and restrained other users' currency liquidity amounting to the maximum of 300 million dollars due to coding vulnerability.

It is the fact that the Blockchain technology is dependent on application software and encoding technologies, and there are quite a few places using algorithms not yet verified among a few hundred startups that develop and provide the Blockchain technology today.

Bruce Schneier, a specialist in encoding and security, has conveyed that the Blockchain network has never been subjected to hacking up to now and it will also remain the same in the future, pointing out "Technology such as Blockchain will not be broken by hacking. Rather, the probability of being broken due to vulnerabilities in software is higher" [26].

Bennet stated "Through cases, it can be seen that there are cases where smart contracts and the concept of manipulation impossibility are incompatible. That is because it means that the worst security bug becomes unrepairable even if it exists, if 'manipulation impossibility' is acknowledged in a smart contract" [26].

Blockchain has the algorithm that cannot be changed once recorded. Lastly, all enterprises using the Blockchain technology should be equipped with common standards and processes. Although there is a movement to solve problems such as R3 consortium working together with the world's largest financial institution, it is a task that is still difficult to resolve.

3.3 Considerations upon the Design of Private Blockchain

When a new system based on private Blockchain is constructed, the items to be considered upon architecture design are as follows.

First, design of the architecture that can guarantee system safety and efficiency as well as maximize advantages of the Blockchain is essential. Secondly, a review of the architecture that can improve update processing speeds for block insertion is required.

Thirdly, structure design that can control the scope of information provision through certification procedure according to user authority should be considered. Fourthly, structure design that allows advance effective checkup of algorithm or vulnerabilities in software should be considered.

4   Study Direction and Results

The present study intends to make a contribution so as to allow safe utilization in P2P distributed network environments by proposing measures for reliable private Blockchain through effective design for the architecture structure of the Blockchain.

Private Blockchain can be divided into 4 elements (Business Network, Asset, Transaction, Smart Contract) that constitute the Blockchain centered on 'Public Transaction Ledger' which all the participants involved in the business can check in real time [9]. In addition, Blockchain technology is to secure the integrity and reliability of transaction records without any authorized third party by verifying, recording and archiving transaction information jointly by all participants in the net-work, and it features a structure that embodies a variety of application services based on distributed network infrastructure by use of security technologies such as hash technology, digital signature, cryptography, consensus, algorithm (PoW, PoS), etc. without certified third parties.

In the end, for a successful business, the Blockchain technology should be prop-erly utilized to build a stable system to ensure that the four elements of Blockchain can function smoothly. Therefore, Blockchain Architecture should be designed effi-ciently considering business characteristics in advance.

As the history repeats itself, the history of the system architecture has been repeat-edly centralized and decentralized, and now it can be deemed that it has entered the era of decentralization again through Blockchain and IOTA, after centralization through Cloud.

As shown in Fig. 4, many social changes are anticipated due to the advent of the new Blockchain technology, which is aimed at decentralization. Especially in the process of designing the system, the architecture design is critical for constructing the whole Blockchain.

From the standpoint of preparing for this social change, businesses based on P2P distributed network are actively expanding on the basis of Private Blockchain, and 4 representative domestic and overseas Blockchains that lead such businesses were identified (Table 4).

Based on white papers, case studies, and press releases concerning each Blockchain, we will identify the outline, architecture structure, major functions and features of Blockchain, and compare the Blockchains to derive improvement points and implications for more effective architecture design.

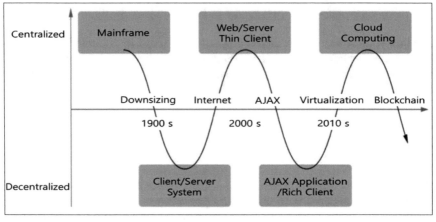

Fig. 4 Diagram of system architecture change

Table 4 Private Blockchain list

Seq	Name	Main group/company	Remarks
1	Hyperledger fabric	IBM	Non-domestic
2	R3 Corda	R3 REV	Non-domestic
3	Nexledger	Samsung SDS	Domestic
4	Loopchain	Theloop	Domestic

4.1 Hyperledger Fabric of IBM

The Linux Foundation's Hyperledger Project begins with 17 members on December 17, 2015 and currently has 150 member companies. It is a collaborative project to develop and implement Blockchain to identify the industry-standard features of the Distributed Ledger that can transform the way business transactions are conducted globally. Hyperledger Fabric is Blockchain Software developed by Hyperledger Project, as an open-source project to develop a framework for decentralized ledgers for enterprise use. It is not a stand-alone project but a project supported by the Linux Foundation for its organization, promotion and technology infrastructure [10].

Blockchain Architecture Fabric 1.0 of Hyperledger optimizes network performance by separating chain code execution and transaction ordering for confidentiality. In addition, it is possible to operate a trusted Blockchain Network based on known participants and regulatory supervision, and to provide multiservice for multiparty transactions through the protection of personal information and confidentiality necessary for regulated industries. In addition, for operational workloads, searchable transaction details can be inquired for efficient audit and dispute resolution, and

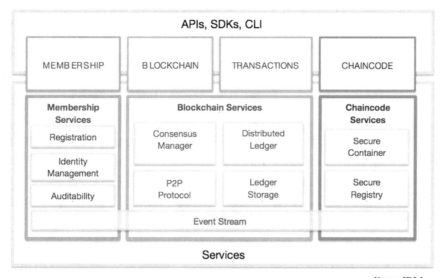

Data : IBM

Fig. 5 The architecture of hyperledger fabric

environment for the number of nodes, agreement algorithm, ID management, and encryption can be set easily for a dynamically expandable business network (Fig. 5).

The Hyperledger Fabric is composed of three major structures. The first one is Membership which identifies the subscription of the node or the identity of the participant. The second one is Blockchain which it implements P2P distributed ledger protocol. And the third one is Chaincode which means ordinary Smart Contract.

In particular, the protocol of agreement of the Fabric uses a modular consensus concept. This is consistent with a well-defined consensus concept in distributed computing, which allows for the development of the Blockchain-related functions of the fabric independently from a specific agreement protocol. The PBFT protocol was implemented as the first protocol in the fabric because of its excellence [11].

Existing Fabric 1.0 integrates many functions in a single node, which may limit achievable performance and damage scalability. Therefore, the improved Fabric 2.0 is presented as shown in Fig. 6.

Fabric 2.0 is an architecture-level approach in terms of scalability and security enhancement and it has improved its ability to run thousands of TPSs to thousands of consortia/colleagues by separating the consensus process [12] (Fig. 7).

The most important feature of the Fabric is network separation, and a new channel (a kind of small Blockchain) is formed according to the type of stakeholders and the application program, and the authority of each node in each channel is distinguished through COP which is an authority management module [13].

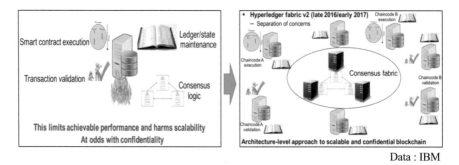

Data : IBM

Fig. 6 Hyperledger Fabric 2.0

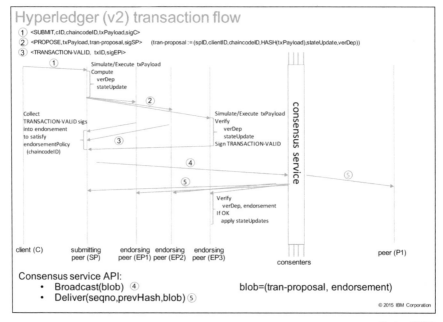

Data : IBM

Fig. 7 Hyperledger Fabric 2.0 transaction flow

4.2 Corda of R3 REV

R3 Corda is a distributed ledger platform developed independently by R3REV, which is a Blockchain consortium consisting of R3, a financial services technology company founded in September 2015. More than 50 global financial institutions such as Goldman Sachs, JP Morgan, Morgan Stanley, and USB are participating in R3 Corda. Founders of R3 identified the most essential problem through a short-term research. That is, the existing Blockchain technologies were not designed to efficiently complement financial institutions' transactions, but were invented by a libertarian motivation of eliminating financial institutions altogether and trading as P2P.

The existing Blockchain code could not be applied to the financial industry merely through the Blockchain code change. For this reason, R3 inevitably decides to build the Architecture layer by layer from scratch. For the first time in financial history, 42 large financial institutions started cooperation and presented the terms of the Internet for financial services that will link them together. In other words, development of Distributed Ledger Platform suitable for the financial industry, which reflects the requirements of customers rather than speculation and assumptions in which started and the Open Source was named Corda. Corda's Architecture is shown in Fig. 8.

Corda is a distributed database technology that equally records/archives the evolution of Contract State (state of the deal) through trading partners' consensus.

The Corda Platform can be considered as a collaborative software platform for financial institutions. Here, collaborative software refers to functions that are jointly processed within the same Distributed Ledger Network. In other words, it is a software for network participants that can jointly process the lower level functions such as the issuance and recording of assets (financial assets such as cash and securities), customer authentication information, time stamping, regulatory reporting, and Oracle (reference data) that are now separately operated by financial institutions.

Beyond a simple mutual data processing, it is to process the operation of a specific part of Finance simultaneously on Network. R3 calls Corda-based financial applications CorDapp. If we compare Corda's architecture with existing Blockchain, Corda creates consensus of a transaction at a deal level between trading partners and divides the consensus into validity and uniqueness. Validity is to confirm that the transaction is Valid when trading partners are satisfied with the content of the transaction and

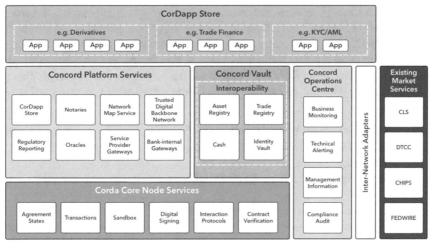

Data: R3 Rev

Fig. 8 The architecture of R3 Corda

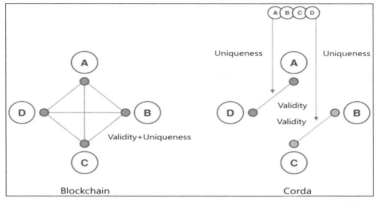

Data : R3 REV

Fig. 9 Consensus mechanism of R3 Corda

they sign. And uniqueness means verification that the transaction has not been spent twice. (*The existing Blockchain does not distinguish the validity and uniqueness of the Consensus. The uniqueness of the transaction is its validity.)

As in Fig. 9, the validity of a transaction is confirmed between trading partners, and the uniqueness of a transaction is guaranteed by introducing the concept of Notary. As the trading partners cannot guarantee that an asset was not spent twice, the Notary steps in and verifies the fact instead. The composition of the Notary may vary from country to country. It can be in the form of a stand-alone institution such as industrial institutions or regulatory/supervisory body such as KFTC or Central Bank, or financial institutions may create a group (pool) to verify uniqueness directly in a decentralized form. If financial institutions create their own Notary group, the group may choose a PBFT or RAFT based consensus algorithm following the choice of the majority. Corda will support a variety of consensus mechanisms [14].

The future roadmap for Corda is to arrive at this design, we first simulated and prototype components of Corda in code to validate aspects of the concept. This is a list of some, but not all, extensions to the Corda model which are expected to be delivered.

- Transaction Decomposition and Uniqueness Enhancements: Incorporating mechanisms to selectively obscure portions of transactions, including obfuscation from uniqueness services.
- Contract Verification Sandbox: Explicit link-time whitelisting of an aggressively minimal set of Java libraries.
- A plug-in based wallet for position inference.
- Calculation oracles or gateways to proprietary (or other) business logic executors (e.g., Central Counterparties or valuation agents) that can be verified on-ledger by participants.
- Using the Corda model to manage identity of users [15].

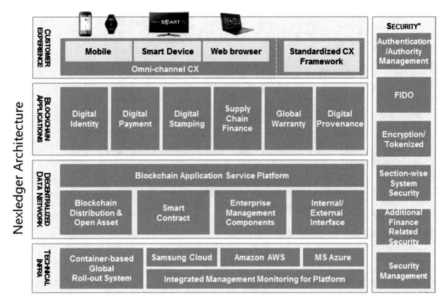

Data: Samsung SDS Nexledger White Paper (2017)

Fig. 10 The architecture of Nexledger

4.3 Nexledger of Samsung SDS

Samsung SDS's Nexledger is a commercialization of Blockchain as a platform service, offering services to domestic and foreign companies and organizations in its main business area, the system integration, SI market (Fig. 10).

Nexledger Architecture is composed of multiple layers. Security layer includes authentication and rights management, encryption and tokenization management while technology Infra Layer includes Samsung Cloud, AWS, MS Azure, and etc. De-centralized network layer for Blockchain distribution and open resource, and Enterprise Resource Management components, Blockchain based application service Layer such as digital identification, digital payment, digital validation, global product guarantee, etc. and customer experience layer such as Omni-channel customer service integrating mobile, web, Smart TV, etc., and standardized customer experience framework are also part of the architecture. The Key Features of Nexledger are, Accounting for the disruptiveness of the nascent Blockchain technology, Samsung SDS created Nexledger to be based on a permissioned Blockchain system and utilize a consortium to fully take advantage of the proven benefits of Blockchain, while eliminating its disadvantages.

The Nexledger platform was developed and reviewed to fit the objectives and requirements of enterprise-level standards:

• Bolstered contingent measures with management monitoring of block information

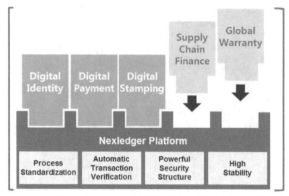

Data: Samsung SDS Nexledger White Paper (2017)

Fig. 11 The architecture of Nexledger platform

- Reduced lead time with improvements in transaction verification and processing algorithm
- Reduced resource consumption with improvements in confirmation racing algorithm for proof-of-work
- Optimized management for the distributed ledger with multi-chains and partitioned chains
- Enhanced security with FIDO certifications and multiple biometric modalities [16].

As the above key features illustrate, Nexledger also made various efforts to improve the consensus algorithm, to enhance security vulnerabilities, and to speed up service processing. Developers who want to implement Blockchain application services, as shown in Fig. 11, will assemble components such as digital identification, digital payment and etc. required for the application as a Lego block [17].

As for the use of Nexledger platform in business, Blockchain can lead to disruptive innovation in international maritime or air logistics transactions, rather than in inland shipping.

As in the case of Fig. 12, management of global supply chain logistics transactions using Blockchain brings additional business opportunities. In other words, adding the conventional maritime-air logistics process to the procurement management process of the global supply chain makes it easier to obtain global logistics visibility [18].

4.4 Loopchain of Theloop

Theloop is a private Blockchain specialist firm started in 2016. The firm independently developed Loopchain, a Blockchain engine, and is currently participating in the financial investment industry consortium as a technology partner.

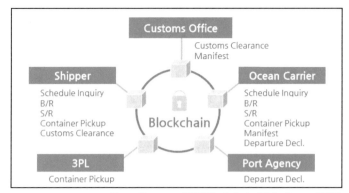

Data: Samsung SDS Blockchain White Paper (2018)

Fig. 12 Samsung SDS maritime logistics platform case (Nexledger)

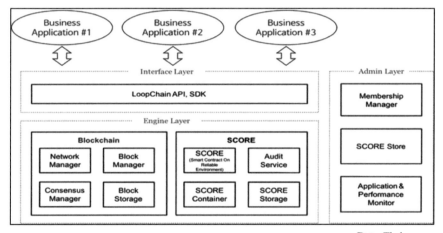

Data: Theloop

Fig. 13 Architecture of loopchain

The financial investment industry consortium is formed by 26 securities companies within the KFIA and 5 IT technology companies, and is the first Blockchain consortium in Korea, created to perform research and development on Blockchain technology and to offer actual services, with the goal of innovating the financial industries. Theloop supports Blockchain engines and solutions in the financial investment industry consortium and has worked to make Loopchain's structure flexible to change the system in order to meet the financial industry's requirements and to meet new requirements (Fig. 13).

In the Loopchain module, the Admin layer mainly manages the failure status of the Node for managing the Blockchain Network, manages the version of Smart Contract (SCORE: Smart Contract on Reliable Environment), supervises the authority of each

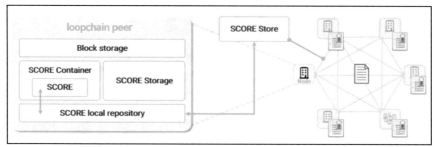

Data: Theloop

Fig. 14 SCORE of loopchain

Node. Engine Layer is responsible for the distributed consensus, the main role of the Blockchain Node, ledger storage, and execution of Smart contract.

In particular, Blockchain, an engine module for distributed consensus, and SCORE, an execution environment, are separated from the services that is loaded on the actual Blockchain. Lastly, the Interface Layer creates an environment allowing other business applications to have access to the Blockchain Network.

The most significant feature of SCORE (Smart Contract On Reliable Environment) is its ability to configure development environment freely through development friendly language. In the case of Ethereum, a representative Blockchain-based Smart Contract Platform, one can only create Smart Contracts in a language available in a special virtual machine, the Ethereum Virtual Machine (EVM). In other words, Smart Contract should be developed through Solidity, Serpent, LLL, and data access and storage can be done only through EVM internal variables [19].

However, as shown in Fig. 14, SCORE is developed as a separate module that minimizes the dependency on consensus engine. The consensus engine and SCORE communicate via the interface implemented by the internal GRPC, so any language can be implemented if the interface is correct. But at this time, only the Python implementation is allowed. The database is also freely available. However, not all data of a variable is stored as it is with Ethereum, and only the results read from and written on the database directly are stored [20].

Private Blockchain technology is taking the center stage, as it allows the customizing of the whole stack of Blockchain, which is required for various transaction validations, external linkage, and regulatory compliance.

Theloop is also exerting the utmost efforts to develop Blockchain, which is more flexible, faster and safer to meet these requirements, and is making improvements.

4.5 Comparison and Implication of Representative Private Blockchain

From the viewpoint of Architecture for Representative Private Blockchains, we summarized the five points of comparison as shown in Table 5: Description of Platform, Governance, Consensus, Smart contracts and Currency.

Through analysis/understanding and cross examination of four representative private Blockchains, the following three implications are derived in terms of efficient architecture design.

First, according to the business model based on a profound understanding on the consensus algorithm, it is important to decide whether to develop a new algorithm or to select and apply an existing algorithm. Second, although the security within the Blockchain is guaranteed, there still is vulnerability in Node and between Node communications, which needs further improvement. Also, the service processing speed needs continued improvement. Third, the various languages used in

Table 5 Comparison of representative private Blockchain

Characteristic	Hyperledger Fabric	R3 Corda	Nexledger	Loopchain
Description of platform	• Modular BlockChain platform	• Specialized distributed ledger platform for financial industry	• Modular BlockChain platform	• Modular BlockChain platform
Governance	• Linux foundation	• R3	• Samsung SDS	• Theloop
Consensus	• Broad understanding of consensus that allows multiple approaches • Transaction level	• Specific understanding of consensus (i.e., notary nodes) • Transaction level	• Broad understanding of consensus that allows multiple approaches • Transaction level	• Specific understanding of consensus (i.e., SCORE, FBFT) • Transaction level
Smart contracts	• Smart contract code (e.g., Go, Java)	• Smart contract code (e.g., Kotlin, Java) • Smart legal contract (legal process)	• Smart contract code (e.g., Solidity) • Based Ethereum (EVM)	• Smart contract code (e.g., PAXOS)
Currency	• None • Tokens via smart contract	• None	• None • Tokens via smart contract	• None

Private Blockchain have not yet matured, and for this reason, Software Bug and vulnerabilities may occur. The use of test techniques to thoroughly inspect software vulnerabilities and preparing an improvement system based on the level of service performance measured are also important factors to consider in architectural design.

Recently, Kim Yong-Dae (2018) emphasized, "Recent research trends include research on source technology for new Blockchain design that guarantees enhanced security, consensus algorithm, and safe equity proof. Nurturing of highly qualified technology human resources in specialized areas including Blockchain security, distributed systems and etc. is urgently needed. Systematic development and supply of Master/Ph.D. degree level experts will contribute to the development of business models that meet market needs and to the activation of business start-ups" [21].

5   Improvement of Effective Private Blockchain Design

We will look at improvement strategies through various related new technologies and case studies in the three areas of consensus algorithm, security vulnerability enhancement, processing speed improvement, S/W vulnerability verification and service performance test enhancement presented above.

5.1   Consensus Algorithm

5.1.1   Definitions and Conditions of the Consensus Algorithm

The consensus algorithm is an algorithm that guarantees the integrity of the system by mutually verifying the mathematically calculated result values through the procedure defined by the nodes that do not trust each other on the distributed network. The consensus algorithm, which resolves consensus problems. There are Paxos and Raft as typical consensus algorithms, and consensus problem is very important in Blockchain. A malicious node in a network and a malfunctioning node are considered spies. In a distributed system that solves the Byzantine general problem, a network can still provide reliable services even if a particular node becomes hacked or malfunctioning. A typical Byzantine Fault Tolerance algorithm is PBFT (Practical Byzantine Fault Tolerance) [22].

5.1.2   Types and Characteristics of the Consensus Algorithm

Private Blockchain uses PBFT (Practical Byzantine Fault Tolerance) and PAXOS algorithms in general, and Quorum, which is a representative project of Enterprise Ethereum, adopts Raft algorithm.

PAXOS selects the leader with the most general consensus algorithm and agrees with majority agreement. PBFT is a consensus algorithm designed to solve the Byzantine general problem and is widely used in Private Blockchain for agreed deduction using 3 step protocols with voting mechanism. Raft is a complement to PAXOS. It is characterized by simplifying the procedure by voting and selecting the leader through random timeout and improving the speed of agreement. There are various algorithms such as SBFT and Tendermint [23].

The Public Blockchain and the Private Blockchain uses different decentralization algorithm due to their network status and digital money operation. In order to use Public Blockchain implementations for the financial sector, it is necessary to replace the algorithm of distributed agreement. Most of the Public Blockchains have a strong dependency on the block processing module and the distributed agreement module, so it is not easy to replace them and there are many problems in future updates. Basically, the Public Blockchain is not flexible to update the software if it is not hard fork once it is published. Therefore, there is a difference from implementing a distributed agreement algorithm in the form of a plugin in the loop chain of Theloop or Hyperledger Fabric which represents the Private Blockchain.

5.1.3   Advances in Consensus Algorithms

A good consensus algorithm means efficiency, safety and convenience, and lots of efforts have been made recently to improve the consensus algorithm. A new consensus algorithm has been devised to solve the specific problem of Blockchain.

Phantom supports accurate and reliable transactions around the world through the Blockchain ecosystem. The OPERA chain, a key technology, is a new distributed Blockchain infrastructure structure that addresses the scalability issues of the existing Blockchain by rapidly processing large blocks.

The Opera Chain is a new Blockchain technology that can be applied to the real world through the support of Smart Contract and story data along with large-scale Blockchain processing. It adopts original Lachesis protocol algorithm which improves limitations of existing agreement protocol such as scalability issues. The Lachesis protocol is a new protocol that dramatically improves the performance and security of the DAG (Directed Acyclic Graph) algorithm used in third generation Blockchain.

The core function of the Lachesis protocol agreement algorithm is to enable fast block processing while safely preserving information and preventing control from specific subjects. It is also possible to completely prevent attacks through hindrance of specific Node (or hacking) such as BFT algorithm (Byzantine Fault Tolerant) [24].

5.2 Strengthen Security Vulnerabilities and Speed up Processing

New technologies are being released each year based on various and continuous researches for enhancing security flaws and processing speed. Among them, new technology that strengthens the security vulnerabilities recently announced to S company is examined, and issues to be considered in designing Blockchain architecture are examined.

SEAL (Secure Encapsulation for Application Layer) is a technology developed to protect data in the data communication section between client and server to counter the security threat caused by OpenSSL vulnerability. SEAL replaces SSL/TSL, a standard communication protocol, and can respond quickly and efficiently to vulnerabilities in communication security based on Open Source. It also suggests that it is enough to replace the existing weak security technologies by solving speed and server/memory resource problems. SEAL is a security technology that protects data among data communication sections in between Client and Server and a technology that is listed in Korea's Information and Communication Technology Association (TTA) in 2015. It is also expected to be listed in the international standard ISO/IEC 11770-4 (Key Management) in 2019. Key features of SEAL include:

- 2-Pass Protocol: Provides fast connection by minimizing the number of message exchanges compared to SSL/TSL
- Connect after Authentication: Prevent security threats such as DDoS attacks through pre-authentication before Client-Server connection
- ID-based Encryption: Provides a simple server authentication system based on ID/PW rather than a certificate.

As shown in Fig. 15, the number of message exchanges is minimized (twice) compared to SSL/TSL (4 times more), and fast connection is provided. As in Fig. 16, SELS's Transaction Per Second (TPS) provides twice the speed of traditional SSL (about 600 times per second) technology.

In addition, as Table 6 shows, SEAL uses 1/20 of memory and 2 times more server utilization (TPS) than OpenSSL. Overall, SEAL has a faster response time and fewer memory/server resources, thus it can be used as an alternative to TLS [25].

Phantom's RAKESIS protocol is fast enough to handle more than 300,000 blocks per second. The Opera chain also uses a quantum cryptography scheme to enhance security when communicating between nodes and a functional programming language for complete Smart Contract support.

Currently, third-generation Blockchain technology has improved performance over previous Blockchain technology, but still has slower block creation and lower transaction performance. The officially announced Third Generation Representative Blockchain's transaction processing per second (TPS) is 258 TPS CARDANO (ADA), 100,000 EOS and STEEM IT 100,000 TPS.

The Phantom Opera chain is an innovative and robust Blockchain technology designed for high production and processing performance of 300,000 TPS and is

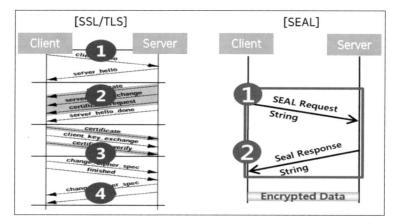

Data: Samsung SDS

Fig. 15 SEAL 2-Pass protocol

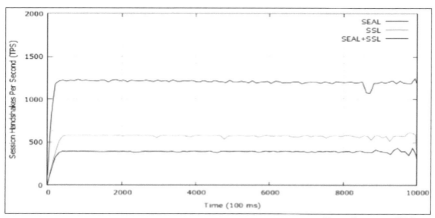

Data: Samsung SDS

Fig. 16 SEAL server performance (speed) versus SSL

available for large scale applications with reliability and scalability. The phantom opera chain is composed of multiple layers to work perfectly in the real world. In the distributed application (DApp) and OPERA Ware Layer, the FANTOM Token is used. In addition, the Opera Core Layer has a unique dual token structure in which an ACTOR, a low-level token, is used [24].

The enhancement of security vulnerability and processing speed are continuously developed, and it is important to explore the application of the Blockchain architecture design based on the understanding of the new technology.

116
K. H. Kwak et al.

Table 6 SEAL versus open SSL technology

	Division	SEAL	Open SSL
Performance	Memory required	2.7 KB/Session	50 KB/Session
	Server performance (TPS)	1,190 TPS	563 TPS
Authorization	Standardization	[Domestic] TTA completed [Overseas] ISO/IEC progress	IETF standard technology

Data: Samsung SDS

5.3 Software Vulnerability Validation and Service Performance Measurement Test

Blockchain technology is still inexperienced and SW is considered to be insufficient, and even the technology of Hyperledger, which is a representative platform of Private Blockchain, is evaluated as inexperienced stage. Thus, this may lead to unexpected problems during the introduction of technology. The various languages currently in use are still not mature enough to be automated, and developers are still learning about these languages. For this reason, there may be bugs or vulnerabilities in the software.

Also, even if the software is developed by strengthening security vulnerabilities, testing and verification methods for the service performance as much as desired in terms of providing the service are still very weak.

5.3.1 S/W Vulnerability Verification

Bennet states that "coding should be able to be modified via an external path in the event of an unexpected situation, such as a mistake in coding". Or at least a final "off-switch (Contract termination method)" should be prepared for an emergency that both parties did not want [26].

As Blockchain is activated in terms of virtual money, the threat of cyber attacks that recently exploited software vulnerabilities has increased and the security of the chain has been secured. However, there are still vulnerabilities to Dapp that constitute them, and the importance of software security is more emphasized due to rising of various new technologies. It is necessary to apply secure coding under the Information Security Management System (ISMS) to implement various applications on Smart Contract using Private Blockchain.

To do this, we utilize an optimal secure coding diagnostic tool that detects and removes security weaknesses in the source code from the development stage. This tool analyzes the source code without actually executing it. It includes a semantic analysis technique for confirming simple grammars and for predicting a value that may be included in advance.

While managing software vulnerabilities that threaten security in the operation of Blockchain can be considered burdensome, such management is essential to reduce security vulnerabilities and may require the use of severity analysis and modification effort evaluation It is expected that efforts will be made easier and cost will be reduced by establishing an improvised plan and managing and implementing the development team by using a task execution tool.

It is important to analyze the severity of vulnerabilities in the event of security vulnerability, and the case studies analyzing seriousness of vulnerabilities using the DREAD of Microsoft Corp at 5 different levels are followed [27].

- Damage Potential: How severe is the damage to the vulnerability exposure?
- Reproducibility: How easily does vulnerability exposure recur?
- Exploitability: How much work has been done to expose vulnerabilities?
- Affected Users: How many users were affected by the vulnerability exposure?
- Discoverability: How easy is it to find vulnerabilities?

After analyzing the severity, infrastructure level vulnerabilities must apply patches that modify the security vulnerabilities that vendors provide, or change configuration settings. However, changing code to address most application vulnerabilities will require a software developer, and it is necessary to determine the level of effort required to actually fix vulnerable code. Based on an understanding of the identified SW vulnerability exposures, the development team determines and orders of modification and what to modify and what not to modify.

Once the plan is established, it can be used to manage SW defects and vulnerability remediation and manage the development team's defect tracking system, and application improvements should be made within the tools that development teams use to manage their business operations.

5.3.2 Service Performance Test

Providing tools to test pre-service performance and determine the integrity of information will be of great help to Blockchain Architecture designers.

Recently, Huawei has developed Service Performance Tool, supported Hyperdger member companies, and has been involved in various development projects [28] (Fig. 17).

The core of this frame is to determine the integrity of the information. For example, it is applied to Smart Contract and checks various ledger book situations. In addition, the stress test of the Blockchain can be performed in a controllable environment, and the result of the transaction success rate, the number of transactions per second, the transaction settlement time, and the CPU consumption amount are generated.

However, as the recent Blockchain competition is getting fierce, it is still uncertain how much this tool will be applied. Like IBM and Intel, Huawei plans to protect Blockchain business performance through patents.

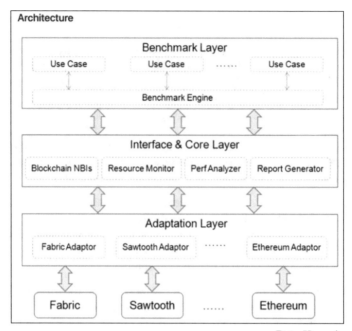

Fig. 17 Huawei Blockchain service performance test tool architecture

Huawei's launch of such tools is also in tune with the telecom industry's competition to develop Blockchain. Recently, Comcast, BT, Telefonica, T-Mobile and other global telecommunication companies such as USA, UK and Spain have been developing Blockchain technology.

5.4 Critical Perspective on Private Blockchain

"Private Blockchain has nothing to do with the Blockchain that people think. It is just a "shared DB" which does not require a proof of work (PoW) and shares a common database in a distributed environment."

That is, in Public Blockchain, the problem of preventing duplicate voting attacks of large anonymous participants is resolved, and in Private Blockchain, the problem of matching DBs between nodes that are already known is solved. Because of this difference, the characteristics of trust and technical utilization value provided by Public and Private Blockchain are also greatly changed.

"The biggest difference between the two networks is the nature of trust," Professor Yang said. "Public Blockchain provides confidence that those who do not participate in the network can trust the chain, while Private or Consortium Blockchain only believed by their participants and the outside cannot believe it."

"It is advantageous (compared to a traditional distributed DB) that an error in a private Blockchain is obtained through consensus within a whole third of the nodes (compared to a traditional distributed DB), but nowadays, it is questionable whether it is appropriate to assume a situation where one third of the attackers will be attacked." "Most of the projects in the recent years are done with mostly Blockchain which is unnecessary," he said [29].

5.5 Application Example and Trend of New Blockchain and Technology

The Cube Chain is a Blockchain Platform with faster, more precise data processing and enhanced security system than the existing Blockchain. The 27 blocks created are formed into one cube, and the cube is connected to another cube to store the data. It is a new concept Blockchain source technology. The Cube Engine is a core technology of the cube chain, and it performs better than the existing Blockchain by forming blocks having three special functions among the 27 blocks constituting the cube by cubing the blocks. In addition, it is an innovative encryption technology that can move data by creating trust in a space where the Internet is not formed by the two-way encryption method through blocking and cubing. At the beginning of cubing, three special blocks are created in the cube. They are called indexing blocks, statistics blocks, and escape blocks, respectively, and have unique characteristics [30].

The Pure Chain develops the Blockchain based on the physical characteristics of the PUF, which quickly creates agreed-upon trading ledgers, does not affect the processing speed, and does not expose unique key values to the outside.

In addition, Pure Chain can be considered as an example of utilizing Blockchain user authentication as a unique attribute in the semiconductor. If the existing Blockchain technologies generate and operate key values based on software, Pure Chain is the main difference from the existing technologies that it is built around hardware technology called PUF to block double payment and forgery risk.

"PUF-based Blockchain technology can be used for existing Blockchain consensus algorithms and cryptocurrency," said Pure Chain. "Pure chains are made up of secure PUF blocks and cannot be tampered with offline. Purechains is the first technology that can be used for the digital money which the central bank is going to issue" [31].

6 Conclusion and Future Research

Recently, Blockchain serves as the core infrastructure that leads the connection with DNA (Data, Network, AI) in Industry 4.0, which is defined as the ultra-connected intelligent revolution based on providing security, transparency, reliability and efficiency. Especially, various application can be implemented based on distributed network utilizing security technology such as hash, digital signature, and encryption.

However, the majority of Blockchain currently has problems such as delays in consensus, delay in processing speed, and user authentication. It is also an issue that cyber attacks are increasing in the P2P distributed network structure. Under these problems, it is required to utilize and continuously research on understanding the agreement algorithm presented in the research result and applying the business model according to the business model, strengthening security weakness in communication between node and node, improving processing speed, and thoroughly checking the vulnerability of S/W The use of test techniques that can be done. Also, continuous improvement is needed to enable customization in case of change.

In addition, in order to activate Smart Block centered Private Blockchain, it is necessary to secure the legal validity of electronic records such as contracts, certificates, and digital signatures registered on the Blockchain Network. Furthermore, it is a time to establish policy, regulation and standardization based on macroscopic perspective analysis. In the field of international standardization, ISO and ITU-T, which are the representative public standardization organizations, have been launched in 2017. The standardization of Blockchain security is currently proceeding with ISO focusing on general Blockchain and ITU-T emphasizing on SG 17.

Blockchain is used in various industries such as digital settlement platform including e-commerce, payment settlement, virtual currency, smart contract used for digital rights and guarantees, stocks and bonds, and cloud funding [30].

This study is limited to empirical research because it is presented as a theoretical case study based on case developed for efficient private Blockchain design. It is considered that the experimental design of the improvement can be a more advanced study in the design of the Blockchain than the actual verification of the improvement through the actual construction.

References

1. Nakamoto, S.: Bitcoin: A Peer-to-Peer Electronic Cash System (2008). bitcoin.org/bitcoin.pdf
2. Compeau, D., Higginsm, C.: Computer self-efficacy: development of a measure and initial test. MIS Q. **19**(2), 189–211 (1995)
3. Bill, S.: Between possibilities and bubbles: 2018 Status diagnosis for Blockchain business. IDG DeepDive (2018)
4. Sehyeon, O., Jongseung, K.: Blockchain nomics. The Korea Economic Daily (2017)
5. Christidis, K., Devetsikiotis, M.: Blockchains and smart contracts for the internet of things. IEEE Access **4**, 2292–2303 (2016)

6. Kim, J.S.: A study on factors affecting accommodation intention of Blockchain technology. Ph.D. degree dissertation at Soongsil University (2016)
7. Jeyeong, L.: Technology trends and implications of Blockchain, Trends and issues (2017)
8. Boucher, P.: How Blockchain technology could change our lives. European Parliament In-Depth Analysis (2017)
9. IBM Korea: Breaking Blockchain for Business. https://blog.naver.com/ibm_korea/221234630201
10. Min-Hee, Y.: A Design and Implementation of Health Insurance Condition Discount System Using Blockchain, Hanyang University Master's thesis (2018)
11. Cachin, C.: Architecture of the hyperledger Blockchain fabric. Workshop on Distributed Cryptocurrencies and Consensus Ledgers (2016)
12. Vukolić, M.: Hyperledger fabric: towards scalable blockchain for business. Technical report, Trust in Digital Life 2016, IBM Research (2016)
13. The Loop: Hyperledger Fabric, R3 Corda (2017). https://blog.theloop.co.kr/2017/02/20/hyperledger-fabric-r3-corda/
14. Baek, J.C.: About the block chain (13) Understanding the Distributed General Corda of R3 pt.2 (2017). http://www.mobiinside.com/kr/2017/06/21/Blockchain-part2/
15. Brown, R.G., Carlyle, J., Grigg, I., Hearn, M.: Corda: An Introduction. R3 CEV (2016)
16. Samsung SDS: Blockchain Platform Case—Samsung Nexledger (2018)
17. Samsung SDS: Blockchain Platform Case—Samsung SDS Nexledger White Paper (2017)
18. Samsung SDS: The logistics revolution that Blockchain will bring. Samsung SDS Nexledger White Paper (2018)
19. The Loop: Loopchain SCORE (Smart Contract On Reliable Environment) (2017). https://blog.theloop.co.kr/2017/04/12/loopchain-scoresmart-contract-on-reliable-environment/
20. ICON Project: ICON Project #6 Inside ICON (3) (2017). https://brunch.co.kr/@helloiconworld/14
21. Shin, Y.O., Kim, Y.D.: Policy debate on Blockchain development (2018) [Online]. http://www.elec4.co.kr/article/articleView.asp?idx=20338
22. The Loop: Consensus Algorithm Overview (2017). https://blog.theloop.co.kr/2017/05/23/
23. The Loop: LFT – loopchain consensus algorithm (2017). https://blog.theloop.co.kr/2017/07/04/lft-loopchain-consensus-algorithm/
24. Lee, J.: 3rd Generation Blockchain 'Phantom Coin' 50 billion ICO promotion (2018). http://www.betaec.net/article/830094
25. Samsung SDS: Secured storage and communications targeted software forms. White Paper (2018)
26. Mearian, L., Bruce Schneier, Bennet: Five Issues of Block Chain. (2017) [Online]. http://www.itworld.co.kr/news/107168?page=0,1#csidxdcdce58c2bcc4efb2eddfd82f2f8f78
27. Software Vulnerability Management Threats to Security: Getting Started with Method—R. https://hmpyun.blog.me/220641011343
28. Yoo, H.C.: Huawei opens the test tool for Blockchain service performance (2018). http://www02.zdnet.co.kr/news/news_view.asp?artice_id=20180316074511
29. Yang, D.H.: NetSec-KR 2018 Topic: Private Blockchain are just Shared DB (2018). http://www.zdnet.co.kr/news/news_view.asp?artice_id=20180418185906&type=det&re==4
30. Nam, S.Y., Kang, M.H., Min, K.S., Lee, K.D., Ang, P.K.: Technology applications of Cube chain and Block chain (2018). http://www.irobotnews.com/news/articleView.html?idxno=13671
31. Son, Y.S.: PUF-based 'Pure Chain' unveiled. In: Delay in Processing Speed (2018). http://www.zdnet.co.kr/news/news_view.asp?artice_id=20180308142218

Designing System Model to Detect Malfunction of Gas Sensor in Laboratory Environment

Ki-Su Yoon, Seoung-Hyeon Lee, Jae-Pil Lee and Jae-Kwang Lee

Abstract Recently, the size of the chemical industry has been growing as well as investing a lot of time and resources to develop core technologies and cultivate human resources. As a result, the handling of chemicals is increasing and the risks are increasing. In particular, there is always the problem of the occurrence of chemical accidents due to the failure of control or management. To prevent this, a disaster detection system using sensors is actively under study. However, the gas sensor among the disaster detection sensors is malfunction due to the influence of the temperature and the humidity. Therefore, in this paper, we collect correlation data of temperature sensor and gas sensor to prevent this. After confirming the correlation through correlation analysis, we calculate regression coefficient by regression analysis and obtain regression equation that can extract sample values of gas sensor data and temperature sensor data. Through this formula, a threshold value that can detect the error value of the gas sensor data is obtained and applied to the decision tree to design a system that can detect the malfunction of the gas sensor according to the temperature change.

Keywords Gas sensor · Regression analysis · Decision tree · Bigdata

K.-S. Yoon · J.-P. Lee · J.-K. Lee (✉)
Hannam University Daejeon, Daejeon, Korea
e-mail: jklee@hnu.kr

K.-S. Yoon
e-mail: ksyoon@netwk.hnu.kr

J.-P. Lee
e-mail: jplee@netwk.hnu.kr

S.-H. Lee
Information Security Research Division, Electronics and Telecommunications
Research Institute, Daejeon, Korea
e-mail: duribun2@etri.re.kr

© Springer International Publishing AG, part of Springer Nature 2019
R. Lee (ed.), *Computational Science/Intelligence & Applied Informatics*, Studies
in Computational Intelligence 787, https://doi.org/10.1007/978-3-319-96806-3_9

1   Introduction

Recently, as the automation technology of industry has developed, the value of simple labor has decreased, and the value of core technology through research has been increasing. Therefore, in order to cultivate the core technology, the state and enterprises invest a lot of time and resources in the development of core technologies and training of researchers [1]. Among them, the chemical industry is directly or indirectly related to almost all industries, and also occupies a very large part in industrial scale. Chemicals are not only raw materials for our daily life but also raw materials that are essential in various high-tech industries such as information technology (IT), bio-technology (BT), and nano-technology or used as a base material [2]. For example, the automobile industry, which is at the center of the machinery industry, is highly dependent on the electronics industry, but now it is increasingly dependent on the chemical industry. Therefore, while the development of the chemical industry and the increase of the chemical circulation have contributed greatly to the development of the national industry, there are many risks, and there is a dark shadow of the risk control and the chemical accident caused by the failure of management. Especially, due to the rapid development of electronics and related industries such as semiconductors, the demand for various new chemicals with high chemical use and harmful risk has increased [4].

Especially, we develop core technologies through various experiments in university research laboratories that have a lot of resources and time. If safety accidents occur, they can lose expensive advanced equipment and major experimental results obtained through much time and effort are lost. Can be. For this reason, safety management of university research laboratories is becoming important.

The more dangerous the chemicals are treated and the more various kinds of materials are used, the more complex the process and the equipment are, and the risk increases doubly if proper process safety management is not done [5]. In addition, chemical plants and chemical handling processes are mostly composed of high-tech complex processes, which are mutually connected and operate in a continuous device industry. In case of a major industrial accident such as fire, explosion or leakage, Not only do they lead to enormous human and material losses, but also pollute the environment or damage local residents [6].

Disaster detection sensors used for this purpose play an essential role in disaster detection. However, since the monitoring system based on the sensor depends on the human resources, the performance of the system may be deteriorated depending on the state of the operator. In addition, occasional sensor faults can lead to malfunctions or improper control behavior by mis-communicating operating conditions or incorrectly performing the control system. However, this method is not able to take immediate action against sensor malfunction. Therefore, it is necessary to study how to continuously grasp the malfunctions and cope with them continuously on-line. This will also be useful in managing and maintaining accidents [7].

Among the disaster detection sensors, gas sensors are influenced by temperature and humidity due to their characteristics, and may be damaged due to malfunction according to the environment. At present, the development of sensors that minimize these effects is ongoing, but it is impossible to develop gas sensors that are fundamentally environmentally unaffected [8].

Therefore, in this paper, statistical analysis is performed using open source R. The correlation between the collected temperature sensor and the gas sensor data is analyzed and regression analysis is applied to confirm the dependency between the data based on the analysis. Through this, we design a decision tree algorithm to investigate the malfunction of the gas sensor. In the case of gas sensors, it is used to detect fire or gas leaks in chemical laboratories. The results of this study can be used to detect gas sensor malfunctions not only in the laboratory but also in other places in the chemical industry.

2   Related Work

2.1   Internet of Things (IoT)

Due to the remarkable development of computing power, sensor nodes with sensing, computing and wireless communication capabilities have emerged in ultra-compact devices. These sensor nodes constitute a mutually autonomous collaborative network and can collect useful information by collecting various information of the surrounding environment. The information collected in such a wireless sensor network is collected from various sensors and utilized in various applications such as an ecosystem environment monitoring system. IoT communication research using MQTT or CoAP protocol is also actively underway [9]. Research on Internet application technology using Raspberry pie and instant detection in case of fire has been going on [10]. In this paper, we study to prevent malfunction of sensor. Collect and analyze gas concentration data as the environment changes. Through this, it is possible to detect the malfunction of the gas sensor data and minimize the safety accidents. In the existing sensor network, each sensor node forms an independent network, so it is not necessary that each sensor node can communicate with the public network. On the other hand, things should be connected to the public network, that is, the Internet. So, in general, keep in mind that the Internet of things has IP stacks up and IP addresses. This allows the Internet to be compatible with all IP protocols and has flexibility and scalability unlike sensor networks. As a result, the Internet is rapidly gaining popularity as a rapidly emerging new field of information generation. It has become important and has expanded rapidly into the field of consumer electronics as well as space, aviation, military and medical fields have. For example, 'smart home technology' from Korea Telecom, 'big data analysis solution for agricultural use' from Fujitsu Japan, and 'sensor-based locomotive fault management' from DB Systel, Germany [11].

2.2 R

It is an object-oriented programming language for statistical analysis and visualization. It is an open source program that has the advantage of adding multiple packages or extending functions. In the analysis of the national climate data using the open source R statistical package, statistical analysis using R was performed to analyze the national climate data [10], but in this paper, the correlation between the gas sensor and the environmental sensor Data analysis was conducted.

2.3 Correlation Regression Analysis

Correlation analysis is used to investigate the correlation between two variables in the data. The Pearson correlation coefficient, which is used to measure the correlation between two variables measured above the isometric scale, is called the correlation coefficient, which is estimated through Eq. (1).

$$r = \frac{\sum_{i=1}^{n} (x_i - \bar{x})(y_i - \bar{y})}{\sqrt{\sum_{i=1}^{n} (x_i - \bar{x})^2} \sqrt{\sum_{i=1}^{n} (y_i - \bar{y})^2}} \tag{1}$$

When the correlation coefficient r is within the range of $-1.0 < r < 0$, it is called a negative correlation and when the correlation coefficient r is within the range of $0 < r < +1.0$, it is called a positive correlation. Figure 1 shows an example of correlation and Pearson correlation coefficient r is defined as Table 1 [12].

Regression analysis is a statistical method for estimating the value of a dependent variable corresponding to a constant value of a new independent variable by finding the regression plane or regression line for the independent variables by measuring the magnitude of the influence of the independent variable on the dependent variable. It is divided into simple regression analysis and multiple regression analysis depending on the number of independent variables. The case where the number of independent variables is one is referred to as simple regression analysis, and the case of two or more independent variables is referred to as multiple regression analysis. The simple regression model is shown in Eq. (2). In this equation, x denotes the independent variable, y denotes the dependent variable, and ε denotes the difference between the value of the actual dependent variable and the value of the predicted dependent variable for a certain independent variable. Because simple linear regression analysis is the goal, residuals are not required [13].

$$y = \beta_0 + \beta_t x_t + \epsilon \tag{2}$$

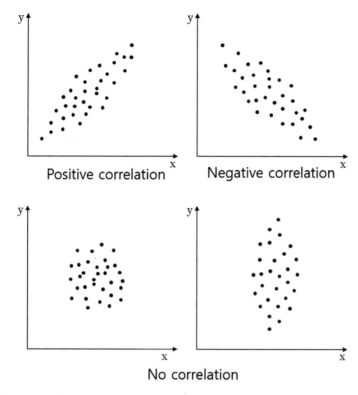

Fig. 1 Correlation diagram

Table 1 Correlation by correlation coefficient

Correlation coefficient range	Relevance	Correlation
$0.7 \leq |r| < 1.0$	Very relevant	Positive correlation
$0.4 \leq |r| < 0.7$	Relevant	
$0.2 \leq |r| < 0.4$	Somewhat relevant	
$-0.2 \leq |r| < 0.2$	Almost irrelevant	–
$-0.4 < |r| \leq 0.2$	Somewhat relevant	Negative correlation
$-0.7 < |r| \leq -0.4$	Relevant	
$-1.0 < |r| \leq -0.7$	Very relevant	

2.4 Decision Tree

Decision trees are one of the most commonly used algorithms in data mining techniques. Decision trees have the advantage that the researcher can easily understand and explain the process as compared with other methods (neural networks, regression analysis, etc.) because the classification or prediction process is indicated by reasoning rules based on the tree structure. In this paper, a decision tree is generated using the C4.5 algorithm proposed by J. Ross Quinlan. The C4.5 algorithm is known to be an algorithm that complements the following problems, which have been pointed out as limitations of ID3 [14].

3   System Model and Methods

3.1   System Model

See Fig. 2.

3.2   Data Collection and Purification

The collected data is in the form of csv and excel files that can be easily analyzed and conveniently found. This data is unprocessed data. In this paper, it is aimed to perform data correlation and regression analysis between gas sensor and temperature

Fig. 2 System model

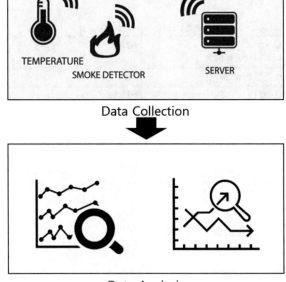

Data Collection

Data Analysis

sensor. Therefore, it is necessary to delete the unnecessary data for the purpose of analysis and align the temperature sensor data with the different gas sensors.

3.3 Correlation Regression Analysis

In order to confirm that the temperature sensor affects the data of the gas sensor, it must be verified whether the gas sensor and the temperature sensor are correlated or not. Therefore, in this paper, the correlation analysis between the temperature sensor and the gas sensor is performed before the regression analysis. To do this, the r coefficient is calculated by substituting the data in Eq. (1). The results are shown in Table 2.

As shown in Table 2, the gas sensor data changes according to the temperature sensor data, and the relationship between the two data can be confirmed through the r coefficient as a negative correlation. This means that as the value of the temperature sensor data increases, the value of the gas sensor data decreases. That is, as the temperature increases, the resistance of the gas sensor that senses the gas increases, and the data value of the gas sensor decreases relatively. Since the absolute value of the r coefficient is 0.7 or more in all four sensors, the correlation between the gas sensor and the temperature is high.

It is confirmed earlier that the gas sensor data value, which is a dependent variable, is affected by the independent temperature sensor data. This time, we try to calculate the regression equation of two data through regression analysis. It is confirmed earlier that the gas sensor data value, which is a dependent variable, is affected by the independent temperature sensor data. This time, we try to calculate the regression equation of two data through regression analysis. The results are shown in Table 3.

Table 2 r coefficient of each sensor according to temperature

	r coefficient
First gas sensor	-0.8120951
Second gas sensor	-0.7953799
Third gas sensor	-0.860681
Fourth gas sensor	-0.8600337

Table 3 Regression equation of temperature sensor and gas sensor and determination coefficient

	Regression equation	Determination coefficient
First gas sensor	$y = -1.82t + 60.66$	0.6595
Second gas sensor	$y = -3.735t + 108.060$	0.6326
Third gas sensor	$y = -4.21t + 120.42$	0.7406
Fourth gas sensor	$y = -4.461t + 128.206$	0.7397

Where t is the value of the independent temperature sensor and y is the value of the dependent gas sensor. In the regression equation of sensor 3, the slope −4.21 indicates the degree of change of the independent variable and the dependent variable by the regression coefficient. The coefficient of determination 0.7406 of Sensor 3 is a measure of how many percent of the sample data used in the regression analysis is explained by the derived regression equation. The determination coefficient of the gas sensor and the temperature sensor is 0.7406, and 74.06% of the sample data can be interpreted as a regression equation. In other words, the correlation between two variables is high and it is confirmed that this regression equation is highly fit.

Figure 3 shows the correlation and the regression equation as a graph, and it is possible to deduce the error value of the gas sensor according to the value of the temperature sensor.

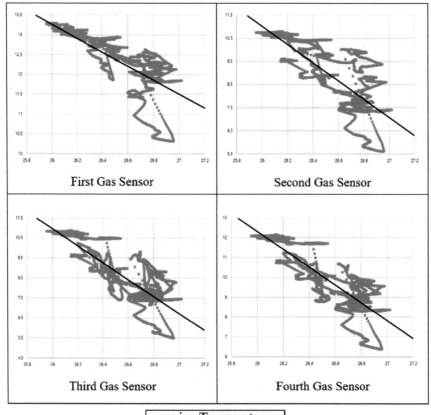

Fig. 3 Correlation and regression equation

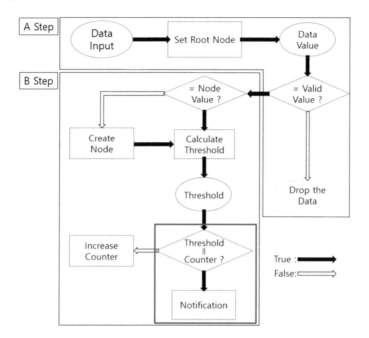

Fig. 4 Decision Tree Algorithm

3.4 Decision Tree

Based on the above correlation and regression equation, we applied the algorithm to the decision tree to design an algorithm that can warn the user when the gas sensor malfunctions. The tree algorithm to recognize the malfunction of the gas sensor is shown in Fig. 4, and A Step and B Step.

First, when the sensor is activated in Step A, the sensor transmits measurement data to the server. If there is no root node, the first received data is set to the data value of the root node. Make sure that the value of the data in the server is valid and store only the valid value. Then, if the data received from the sensor is different from the data value of the root node in step B, the data value of the root node is compared with the data value of the root node to set the data of the new child node and calculate the threshold value for the node. If the new data is the same as the data of the terminal node, the algorithm is designed to increase the counter value and autonomously recognize the malfunction if the counter value is above the threshold value. Here, the threshold value applied according to the sensor data uses the given value through the regression equation, and it is possible to correct the error value of the gas sensor according to the temperature sensor.

4 Conclusion and Discussion

In this paper, we design a model to propagate behavioral rules to prevent the spread of secondary damage by autonomously recognizing laboratory safety accidents. We constructed a system for data collection and analysis, and proposed a decision tree algorithm suitable for recognizing laboratory accidents. Based on the data collected by the smart sensor, the researcher can autonomously recognize, analyze, process, predict the secondary damage, and design a model that can propagate the behavior. This enables systematic management of safety accidents in the laboratory and prevents unnecessary secondary damage, minimizing losses in the laboratory.

In this paper, data analysis is performed to detect malfunction of gas sensor. Data were collected from data collected in an environment with the same gas concentration and only a change in temperature. As a result of the experiment, it was confirmed that the gas concentration data change according to the change of the environment even though the same gas concentration. Therefore, correlation analysis was performed to analyze the correlation between data.

As a result of correlation analysis with temperature sensor data, the absolute correlation coefficients of all four gas sensors were 0.7 or more and the average value was -0.8320292. This indicates that they have a high correlation and a negative correlation. Therefore, the higher the temperature, the more likely it would have a negative impact on the gas sensor data.

Based on the correlation analysis, the regression equation and the coefficient of determination were calculated by regression analysis of the temperature sensor and the gas sensor data. As a result, the regression equation showed that the sample data can be interpreted with an average of about 69.25%. Based on these results, it can be expected that the sensitivity of sensor data can be minimized by correcting the value of gas sensor data by applying it to the decision tree algorithm.

However, considering the average value of the regression equation, it is difficult to detect the malfunction of the gas sensor perfectly because the data of the temperature sensor can be interpreted only about 70% of the data of the gas sensor. In the future, we will carry out research to detect the malfunction of the gas sensor by using the data of the humidity sensor as well as the temperature sensor. And we study algorithms that increase the accuracy of gas sensor measurement data based on the regression equation of integrated data. By applying this to the decision tree, it is expected that the false alarm due to the malfunction of the gas sensor will be reduced, and more precise gas detection will be possible.

Acknowledgment This research was supported by Basic Science Research Program through the National Research Foundation of Korea (NRF) funded by the Ministry of Education (NRF-2017R1D1A3B03036130)

References

1. Kim, J.S.: A big data based spatio-temporal sensor data processing system. Unpublished doctoral Dissertation, Konkuk University (2017)
2. Ha, D.M., Lee, S.J., Park, S.H., Jang, S.Y., Han, J.S., Park, J.H., Byun, M.J.: The handling standard of toxic gases for laboratory safety. J. Korean Inst. Gas, 69 (2014)
3. Kim, S.C., Song, Y.H.: A study on the building plan of chemical management system in laboratory using FGI method. Fire Sci. Eng. **31**(1), 18–25 (2017)
4. Eom, S.H., Lee, S.K.: A study on analysis of laboratory accident with root cause analysis. J. Korean Inst. Gas **14**(4), 1–5 (2010)
5. Ha, D.M., Lee, S.J.: A safety management of chemical and chemistry engineering laboratory for fire prevention. Fire Sci. Eng. 114 (2013)
6. Jang, S.I., Shin, D.M., Kim, T.O., Park, Kyoshik: Development of a guideline for accident investigation and management for an experimental laboratory. Korean J. Hazard. Mater. **4**(1), 57–64 (2016)
7. Lee, K., Kim, H.G., Lee, B.W., Kim, T.O., Shin, D.: Test bed design of fire detection system based on multi-sensor information for reduction of false alarms. J. Korean Inst. Gas **16**(6) (2012)
8. Deschambault, O., Gherbi, A., Légaré, C.: Efficient implementation of the MQTT protocol for embedded systems. J. Inf. Process. Syst. **13**(1), 26–39 (2017)
9. Lee, Y.M., Sohn, K.R.: Fabrioation of smart alarm service system using a tiny flame detection sensor based on a Raspberry Pi. J. Korean Soc. Mar. Eng. **39**(9), 953–958 (2015)
10. Kim, D.S., Lee, J.W., Kim, K.H., Kim, H.S., Cho, S.M., Hyun, M.C., Lee, S.E., Cho, S.I.: Analyze national climate data using open source R statistical package. In: Proceedings of the Korean Meteorological Society Spring Conference, pp. 76–77 (2014)
11. National Information Society Agency ICT Convergence: 2016 Global Big Data Fusion Casebook (2016)
12. Hong, S.K., Kim, S.P., Lim, S.H., Moon, Y.S.: Secure multi-party computation of pearson correlation coefficients. J. Korea Inf. Sci. Soc. 171–173 (2014)
13. Yoo, C.S., Park, C.S., Yoon, J.S., Ha, E.H.: Regression analysis or the log-normally distributed data and mean field bias correction of radar rainfall. J. Korean Soc. Civ. Eng. B **31**(5A), 431–438 (2011)
14. Quinlan, J.R.: C4.5: programs for machine learning. Morgan Kaufmann Publishers, San Mateo, Burlington, Massachusetts (1993)

Design of Device Mutual Authentication Protocol in Smart Home Environment

Jae-Pil Lee, Seoung-Hyeon Lee, Jae-Gwang Lee and Jae-Kwang Lee

Abstract Amid the rapid expansion of smart devices including smartphone in our daily lives, more and more households have been adopting new forms of Smart Home services by linking home electronics with their smart devices. This kind of Smart Home services are aimed to make people's personal spaces more comfortable and convenient. However, the connection among many smart devices could lead to a situation in which the devices could be controlled by third parties through malware, unauthenticated access or hacking, increasing the risk of device modulation, privacy invasion, and date leak. Therefore, it is imperative to come up with some security requirements. There is a problem of the existing home network system: it allows multiple ways of protocol, making it impossible for devices with different protocols to be connected and provide safe services. Also, the whole system would stop if a system component does not work in the server-based platform. This paper will suggest a security protocol for device/user authentication and access control in order to enable easy and convenient compatibility services. In addition, it gives you a way to establish a safe and secure framework thorough the protocol suggested.

Keywords Smart home · Block chain · Mutual authentication · Devices

J.-P. Lee · J.-G. Lee · J.-K. Lee (✉)
Department of Computer Engineering, Hannam University, Daejeon, South Korea
e-mail: jklee@hnu.kr

J.-P. Lee
e-mail: smartfeel@gmail.com

J.-G. Lee
e-mail: jglee@netwk.hnu.kr

S.-H. Lee
Information Security Research Division, ETRI, Daejeon, South Korea
e-mail: duribun2@gmail.com

© Springer International Publishing AG, part of Springer Nature 2019
R. Lee (ed.), *Computational Science/Intelligence & Applied Informatics*, Studies
in Computational Intelligence 787, https://doi.org/10.1007/978-3-319-96806-3_10

1 Introduction

Smart Home refers to a space that integrates Information and Communication Technology (ICT) into a residential environment and provides a variety of information and provides a variety of information and services without restricting space and equipment. This enables economic benefits, health, welfare, increased quality of life [1]. As for Smart Home, home appliances such as a TV, a refrigerator, and a washing machine are connected to a wired/wireless home network. Therefore, regardless of the user's device, time, and place, convenient and safe environments are established and people can now enjoy various Smart Home services. According to Strategy Analytics, a global research organization, the market expansion is attributable to the development of IoT (Internet of Things) services and smart devices and the increase in demand for Smart Home services. As for 2014, the global Smart Home market is estimated at $48 billion, growing at an average annual rate of 18.4% starting in 2014 and reaching $111.5 billion by 2019 [2]. With the rapid expansion of smartphones and smart devices in recent years, many households are adopting a smartphone-based computing environment instead of a PC-based computing environment, which is bringing about a paradigm shift in our daily lives. As information appliances and white goods evolve into Smart Home appliances, various wearable appliances are entering the home network environment. New types of Smart Home service have emerged, constructing a safe and comfortable space for people. For example, by utilizing IoT technology, we are trying to form a cooperative relationship between existing home appliances and smart devices.

However, as the number of devices connected to smart appliances exponentially increased, the devices could be connected to the Internet and controlled by third parties. Therefore, it causes problems such as increase of security threat, unauthenticated user access, forgery of device due to malicious code hacking, and this may lead to personal privacy invasion problem. Smart Home environment is resource-limited because IoT home appliances Smart Home appliances such as smart TV, smart refrigerator and IoT household appliances and sensors use low-power, low-end CPUs. Also, they are only available through heterogeneous sensors. IoT standard platforms have been developed and become applicable to Smart Home. Prominent examples of standard platforms are Open Connectivity Foundation (OCF), Thread Group, HomeKit, OneM2M, and Industrial Internet Consortium (IIC). Today, global IT companies and operators are working to spearhead global standards and working on platform development and Smart Home services [3, 4]. For now, there is no standard technology that leads the Smart Home market, and many businesses are actively developing standardization technology to lead the global IoT industry. As a result, security threats in the existing network environment are increasing because Smart Home services are provided in a network environment in which various kinds of devices are integrated [5]. For example, safety incidents have occurred several times in unsecured Smart Homes, which suggests that lightweight devices with poor security are one of the biggest problems. Therefore, in order to operate the Smart Home service safely, it is necessary to make sure light devices connected to the Smart

Home are safe and secure. The Smart Home environment minimizes user intervention so that home devices share and be connected to each other. Accordingly, in order to provide connectivity and compatibility between the devices, Smart Home should have an authentication system. The existing home networks have compatibility problem when using different protocols using various protocols, and often fail to provide secure services. In addition, the existing ID-based, server-centric platforms have a serious problem: if the system components do not work, the entire system would stop.

In this paper, we propose a security protocol that enables the authentication access control of users and devices. In addition, we propose a secure Smart Home protocol design. The contents of this paper are as follows. In Sect. 2, we will discuss security requirements and device authentication technologies required in Smart Home. Also, USIM, token, block chain, and ACL information are described in details. In Sect. 3, we propose a mutual authentication framework for Smart Home devices. A protocol for authentication token, block chain, and policy settings will be constructed and a mutual authentication protocol based on Smart Home services will be proposed. In Sect. 4, we will move on to establish the environment for testing the proposed device mutual authentication. In Sect. 5, we will conclude our research and discuss the subjects and fields for further research.

2 Related Works

2.1 Security Threats of Smart Home

Smart Home appliance devices are connected to each other through a network and share information, making users' lives comfortable and convenient. However, security threats exist. For example, thief or loss of information devices, IP spoofing, DoS attacks, Trojan horses, worm viruses, and traffic analysis can happen, resulting in security threats such as data modulation, illegal authentication, and privacy invasion [6]. IoT products used in Smart Homes require computing power and network connectivity. There are various security threats because it collects and manages the data of devices of various home appliances. Unlike general ICT systems, Smart Home products are relatively insecure because they are difficult to apply security technology. In addition, home appliance devices used in networks have various computing capabilities, and devices with low computing power are likely to be targets of cyberattacks due to the lack of security functions. As for Smart Home devices, there are many devices that are used without proper security measures. unauthenticated smart devices are easily accessible to the devices through the network. Therefore, within the home gateway that contains sensitive information of the user, it is necessary to control unauthenticated devices. Sensitive information including personal information, device settings, messages about the terms and conditions are transmitted through the network. Therefore, in a Smart Home environment, data must be encrypted when they are transmitted, so that hackers or unauthenticated users cannot access. In a Smart Home environment, a user can send or receive data in and out through a wireless network. Therefore, in a Smart Home environment, data stored and exchanged between users and devices should be protected from forgery. Security technologies such as authentication, confidentiality, integrity, availability, access control, and denial of access are required. In terms of services, authentication, encryption, and data protection are required [7].

2.2 Device Authentication Technology

In order to set up the device authentication technology in the Smart Home environment, the authenticated smart device should be registered in the vicinity of the home gateway, and communication should remain safe and secure through the authentication process. Since there are various security threats, countermeasures are necessary, and it is necessary to securely manage and authentication devices. The devices used in the Smart Home environment have a disadvantage, given it is difficult to accurately authenticate and identify the device without direct intervention of the user. In general, ID/PW, MAC address, USIM, and password-based authentication technology are used, and temporary password-based authentication technology is also used depending on service environment [8].

The ID/PW authentication technique is very convenient and relatively easy to implement compared to other technologies. However, in the Smart Home environment, authentication level is low, making it easy for an attacker to access. Home appliance devices communicate in a TCP/IP environment. The MAC (Media Access Control) is a hardware address and is used in a network environment. MAC address-based authentication is faster and more convenient than ID/PW method, but vulnerable to forgery in the Smart Home environment.

The cryptographic-based technology used to authenticate Smart Home devices includes a symmetric-key cryptographic algorithm and a public key algorithm classified according to the type of the key used for encryption and decryption. For the two-way authentication between the devices, protocols or encryption algorithms used for device authentication for wireless settings are mainly used, such as 802.1x, Wi-Fi Protected Access (WPA), Extensible authentication Protocol (EAP), and RFC3748. Compared to ID/PW, and MAC address-based authentication, its security level is fairly high. However, there exists a certain level of vulnerability according to the protocol and algorithm used.

The symmetric key encryption algorithm is constructed as follows. The transmitted data, which is a plain text, produce cipher text by using the algorithms of symmetric key (secret key) and encryption and the resulting cipher text is transmitted to the receiver. The user that received cipher text generates a plain text by using the same symmetric key and decryption algorithm. In the symmetric key cryptosystem, the encryption key and the decryption key are the same in order to encrypt the device authentication information. As the number of other parties to transmit and share data through encryption and decryption increases, the number of shared secret keys increases. Therefore, it is necessary to keep the keys confidential.

The public key cryptosystem, which is another cryptosystem, is based on a difficult mathematical problem such as a factorization problem and a discrete algebra problem. Therefore, even if a key is disclosed as a public key, it is difficult to find the pair of the private key. The basic encryption scheme for maintaining the confidentiality of data uses the recipient's private key to encrypt the data transmitted by the sender. The user who receives the cipher text decrypts it with their private key and obtains the plain text. In case of a third-party attack, even if the transmitted cipher text is acquired on the network, it cannot be decrypted with plain text until the key used for decryption is obtained. However, there is a possibility of attacking the middle man by eavesdropping at the process of exchanging the identifier between the Smart Home authentication server and the client device.

A Smart Home appliance generates a One-Time password that can only be used in the session that is connected to the Smart Home gateway. The most commonly used method is a method of authenticating users with random temporary passwords by using One-Time Password (OTP). As Smart Home appliances are mostly light and low-powered, there is a limitation in applying the existing authentication system to the Smart Home environment.

The universal subscriber identity module (USIM) is a combination of a SIM card with subscriber information and a universal IC card (UICC) used for user authentication, global roaming, and E-commerce. USIM is composed of a CPU with encryption and decryption functions and a memory for storing additional service functions such as transportation cards and credit cards. However, there exist security concerns, for example, if a locked device is lost, personal information can be leaked. A related study that seeks to solve the problem of authentication and device authentication for legitimate users in the smart work environment can be found at [9] The study suggests some solutions for device loss, malicious code and leakage of corporate confidential information. This study utilizes the USIM information and device authentication ID to block unauthenticated users and prevent important data from leakage.

2.3 Block Chain Technology

Block Chain has emerged as a new paradigm shifting the existing business flow mainly in the financial sector. It is also considered a strategic technology for the future [10]. Block Chain, introduced by Satoshi Nakamoto, the founder of Bitcoin [11], is a system that prevents information from being modulated and shares it transparently and securely. Bitcoin started its service in 2009, and now the technology has been fully commercialized. Unlike the pre-existing currencies, Bitcoin does not have a central authentication to manage participants, and all participants are connected based on P2P network. Block Chain technologies include peer-to-peer networks, cryptography, decentralized ledger, decentralized contracts, and Smart Contract-based technologies [12]. Instead of the pre-existing centralized server management method, a user participating in the transaction distributes and saves a collection of data. Since it is confirmed continuously whether or not the blocks are consistent with the blocks of each user, it is on its way to becoming a key technology to secure data reliability and confidentiality. It also is evolving into logistics, pushing the boundaries of other sectors as well. Smart Contract is a type of contract that was first introduced in 1994 by the cryptographer Nick Szabo and integrated into Block Chain. If entities or objects agreed on certain conditions and terms, transactions are automatically made or payment are completed. Smart Contract is applied to the block chain technology with time stamps, in order to certify ownership, automobile, housing/real estate contracts, and copyright. Smart Contract minimized the number of brokers for reliability in transactions and minimize harmful exceptions, also meeting the contract terms at the same time. Block is a container data structure formed to include transactions in a block chain. A data structure is a list of blocks in which transactions are associated with the previous block. Blocks within a block chain are identified using hash data. The hash data is generated using the SHA256 cryptographic hash algorithm in the header of the block. The block has a long list of headers containing metadata and a transaction list that determines the size of each block [13]. The structure and operation of the block chain include blocks of the previous block, the current transaction information, and the hash value. Therefore, the block structure is formed by associating the block of the transaction information with the previous block. And it is difficult to modulate the contents. In addition, given the transaction information is disclosed, it can be managed transparently.

3 Design of Device Mutual Authenticaion

In this chapter, we design the authentication system that can share and use authentication information of users and devices in a safe manner.

The system will solve the problem of the pre-existing centralized home server using the authentication token and the user ID of the Smart Home user. A Smart Home user can receive services for every device simply by completing the initial registration process without any additional registration processes. Figure 1 shows the outline of the device mutual authentication in the Smart Home environment. It consists of devices, Home Gateway, and Block Chain.

The device mutual authentication protocol proposed in this paper is an ID connection method that provides mutual authentication services between device devices by querying token information registered in a block chain using a user ID in the Smart Home environment. As shown in Fig. 2, the system diagram consists of USIM, key paring, encryption/decryption, and nonce generation modules.

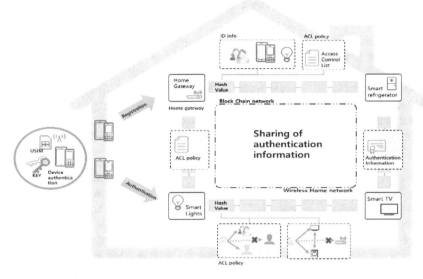

Fig. 1 Overview of mutual authentication for smart home devices

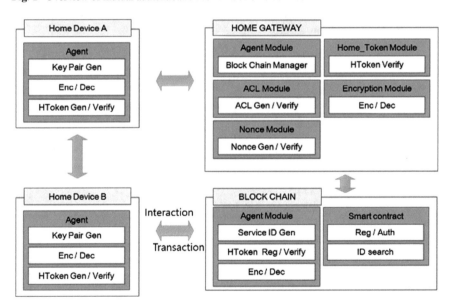

Fig. 2 Smart home system configuration

3.1 Device Registration Process

The registration process for the Smart Home environment is Fig. 3. First, you need to register for Smart Gateway (SGW) by using the device information with the installed USIM card. The registered user and device information is used as authentication information for the mutual device authentication in the Smart Home network environment in SGW. In order to use the Smart Home authentication service, it is necessary to set up the agent of the new SGW settings. It is also assumed in this process that the user confirmation is completed through the USIM card. The following describes the device registration process.

Step 1:
To identify the user/device and register the token, you have to hash the user ID, the device public key, the S/N, and MAC information to the device's public key. And send the encrypted private key to the SGW along with the public key. The SGW decrypts the encrypted information using the public key received from the device and authenticates the public key by comparing the hash result with the public key. Register the device in the SGW afterwards. In the SGW, a nonce generation module is used to generate a nonce value by combining the MAC, which is the original information of the SGW, with the device public key and the user ID information, which are the user's information.

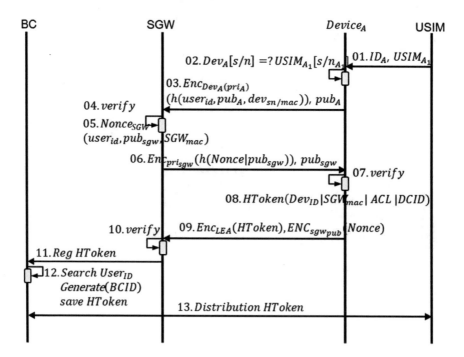

Fig. 3 Device registration process

Step 2:
Hash the public key information of the SGW by using the public key and send it to the device along with the public key of the SGW. Then, the device confirms the nonce value transmitted from the SGW and generates a DCID and a home token (HToken) for device verification with the nonce value. The home token is encrypted using the Lightweight Encryption Algorithm (LEA) encryption algorithm, and the key value is encrypted with the public key of the SGW and transmitted to the SGW. The SGW decodes the received value from the device and verifies whether it is the value generated by the SGW. The home token value is verified by decrypting the encrypted token using the LEA with the proved nonce value as a key.

Step 3:
After the verification process, create a block based on the home token and public key information generated by the SGW using the block chain service ID (BCID) accessible from the block chain using the user ID information. Then distribute the block chain information to the devices registered through the SGW.

3.2 Device Authentication Process

Figure 4 shows the mutual authentication process of querying and authenticating the information of the token registered in the block based on the user ID in the device.

Step 1:
During the process of device authentication, confirm the user ID registered in the SGW using the user ID information available at SGW. Then, the SGW confirms the registered user and device identification by checking the user ID information. After that, the SGW confirms the block chain service ID list registered in the block chain based on the user ID, encrypts the BCID information matched with the user ID, and transmits the encrypted BCID information to the device through the SGW.

Step 2:
Device A decodes the BCID information received from the block chain and confirms the BCID. Also, Device A generates a device home token and encrypt it with the BCID to request a home key token verification with a block chain. The block chain compares the home token information registered in the block with the home token information received from the device A to verify the information of the home token. As a result, if the verification is normally performed, the authentication result is transmitted to the device through the SGW. The SGW then generates a session key, encrypts it, and sends it to the device. Device A uses the received session key and connects the session.

Step 3:
Device B requests device A to access the service. Device A inquires the access right of device B using the home token received in the block chain. After the token verification

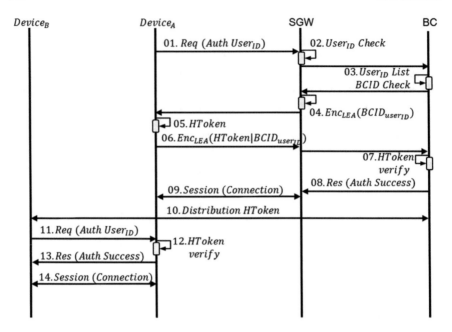

Fig. 4 Device authentication process

is completed, the device A transmits the authentication result to the device B. Then device B can use the Smart Home service.

4 Mutual Authentication Experiment

In this chapter, a block chain network is constructed by using the architecture of Fabric v1.0 based on device mutual authentication design.

In order to construct a sample network, it is assumed that each peer has an ordering service that can consistently manage the same ledger for transactions, and four peers (devices) that conduct the transaction and manage the ledger. Table 1 shows the development and experiment settings.

As for the scenario, we constructed the block chain network, set up the initial device policies, and distributed them. And, we checked the authentication information by querying the policies set in the devices.

Figure 5 shows the process of generating a genesis block file using the user's block chain ID (home). Create a channel using the user's block chain ID and set it for each channel so that the consumer devices can enter the block-chain network When all appliances enter the network, you need to use the chain code to set the device policy and distribute the initial chain code, use "invoke" for data creation, and use "query" to search. When all appliances enter the network, you need to use the chain code

Table 1 Experimental settings

Host OS	OS	Windows 10 Home
	Web browser	[Google Chrome] 64.0.3282.186 (64-bit)
	JavaScript library	[jquerty.min.js] 2.1.0
	Develop tools	[IntelliJ IEDA Community Edition 2017.3.2]
	Virtual S/W	[Virtual Box] 5.1.30
	Etc tools	[Git Client] 2.9.2 [Go] 1.8 [Vagrant] 1.9.2 [PuTTY] 0.66
Guest OS	OS	[Ubuntu Server] 16.04.4 TLS
	BlockChain	[Hyperledger Fabric] 1.0.0
	Etc tools	[Go]1.6.2 [Docker] 1.110.2-0

```
2018-03-21 16:02:59.527 UTC [msp/identity] Sign -> DEBU 01e Sign: plaintext: 0ABE06C
0522...8EDE3B82ECAE12080A021A0012021A00
2018-03-21 16:02:59.527 UTC [msp/identity] Sign -> DEBU 01f Sign: digest: 3F6629BF9C
A3ECA8178841D22B9E988BF9DF1E39284
2018-03-21 16:02:59.533 UTC [channelCmd] readBlock -> DEBU 020 Received block:0
2018-03-21 16:02:59.534 UTC [main] main -> INFO 021 Exiting.....
root@4a5853cfeac7:/opt/gopath/src/github.com/hyperledger/fabric/peer# ls
channel-artifacts  crypto  home.block  scripts
root@4a5853cfeac7:/opt/gopath/src/github.com/hyperledger/fabric/peer# 

2018-03-21 16:07:32.048 UTC [golang-platform] GetDeploymentPayload -> DEBU 00a done
2018-03-21 16:07:32.050 UTC [msp/identity] Sign -> DEBU 00b Sign: plaintext: 0A89070
0510...5F74FD270000FFFFCEF44F9B002C0000
2018-03-21 16:07:32.050 UTC [msp/identity] Sign -> DEBU 00c Sign: digest: F54FBD951E
88A4A4409876ECA519CEF56CEDE4F6FF6
2018-03-21 16:07:32.058 UTC [chaincodeCmd] install -> DEBU 00d Installed remotely re
yload:"OK" >
2018-03-21 16:07:32.058 UTC [main] main -> INFO 00e Exiting.....
root@4a5853cfeac7:/opt/gopath/src/github.com/hyperledger/fabric/peer# 
```

Fig. 5 Generating a block file

to set the device policy and distribute the initial chain code, use "invoke" for data creation, and use "query" to search. The contents of the device policy are a list of devices that entered the channel. You can set the policy to verify the transactions for certain devices. Then, the devices will receive transaction offers.

See Fig. 6. If a block-chain network receives a signature from one of the devices containing the transaction, it determines that it is a legitimate transaction. As for the initial device policy, the authentication value of the device A is set to "7 (read, write, execute)" and the authentication value of the device C is set to "4 (read)". Figure 6 shows the policy setting process of the device.

Figure 7 shows the process of inquiring the authentication policy set in the devices. After the device policy is set up as shown in Fig. 6, the policy information of the device A in the Home block can be inquired through "Query". In Fig. 6, when a transaction is conducted through "Invoke", the policy information set in Device A can be checked. However, you should check if the other device is working properly and that the ledger is properly distributed.

```
root@4a5853cfeac7:/opt/gopath/src/github.com/hyperledger/fabric/peer#
nization/org1.example.com/users/Admin@org1.example.com/msp/peer/crypto/peerOrga
ADDRESS=peer0.org1.example.com:7051ithub.com/hyperledger/fabric/peer# CORE_PEER_
LOCALMSPID="Org1MSP"pt/gopath/src/github.com/hyperledger/fabric/peer# CORE_PEER_
Organizations/org1.example.com/peers/peer0.org1.example.com/tls/ca.crt/crypto/peer
 "OR ('Org1MSP.member','Org2MSP.member')"":["init" "DevA", "7", "DevC","4"]' -P
2018-03-21 16:21:16.606 UTC [msp] GetLocalMSP -> DEBU 001 Returning existing local M
2018-03-21 16:21:16.606 UTC [msp] GetDefaultSigningIdentity -> DEBU 002 Obtaining de
y
2018-03-21 16:21:16.610 UTC [chaincodeCmd] checkChaincodeCmdParams -> INFO 003 Using
2018-03-21 16:21:16.610 UTC [chaincodeCmd] checkChaincodeCmdParams -> INFO 004 Using
2018-03-21 16:21:16.612 UTC [msp/identity] Sign -> DEBU 005 Sign: plaintext: 0A90070
0510...324D53500A04657363630A0476736363
2018-03-21 16:21:16.612 UTC [msp/identity] Sign -> DEBU 006 Sign: digest: B5A1A1B43E
54AD89BBC9A6969AE287B8CBDC70E9C3A
2018-03-21 16:21:34.709 UTC [msp/identity] Sign -> DEBU 007 Sign: plaintext: 0A90070
0510...5205B7E8EABFDD32FA233634F53C417F
2018-03-21 16:21:34.709 UTC [msp/identity] Sign -> DEBU 008 Sign: digest: 296E55638B
0B922E5DB153CF64FE51726C160A62A14
2018-03-21 16:21:34.714 UTC [main] main -> INFO 009 Exiting.....
root@4a5853cfeac7:/opt/gopath/src/github.com/hyperledger/fabric/peer# □
```

Fig. 6 Setting device policies

In Fig. 7, it distributes the device policies and executes "Query" for the device C. If you query the policy value of device A in the device C after executing the query command, the policy value "7" is revealed. The ledger will keep getting updated

```
root@4a5853cfeac7:/opt/gopath/src/github.com/hyperledger/fabric/peer#
root@4a5853cfeac7:/opt/gopath/src/github.com/hyperledger/fabric/peer#
nization/org1.example.com/users/Admin@org1.example.com/msp/peer/crypto,
ADDRESS=peer0.org1.example.com:7051ithub.com/hyperledger/fabric/peer# CC
LOCALMSPID="Org1MSP"pt/gopath/src/github.com/hyperledger/fabric/peer# CC
Organizations/org1.example.com/peers/peer0.org1.example.com/tls/ca.crt/peer
code query -C home -n HomeACL -c '{"Args":["query","DevA"]}'ric/peer# pe
2018-03-21 16:23:30.558 UTC [msp] GetLocalMSP -> DEBU 001 Returning exis
2018-03-21 16:23:30.558 UTC [msp] GetDefaultSigningIdentity -> DEBU 002
y
2018-03-21 16:23:30.558 UTC [chaincodeCmd] checkChaincodeCmdParams -> IN
2018-03-21 16:23:30.558 UTC [chaincodeCmd] checkChaincodeCmdParams -> IN
2018-03-21 16:23:30.558 UTC [msp/identity] Sign -> DEBU 005 Sign: plaint
0510...4C1A0D0A0571756572790A0444657641
2018-03-21 16:23:30.558 UTC [msp/identity] Sign -> DEBU 006 Sign: digest
11CFE5462C7C1C82359E3D0A9B3125E10A
Query Result: 7
2018-03-21 16:23:30.572 UTC [main] main -> INFO 007 Exiting.....
root@4a5853cfeac7:/opt/gopath/src/github.com/hyperledger/fabric/peer# ▮

root@4a5853cfeac7:/opt/gopath/src/github.com/hyperledger/fabric/peer#
nization/org3.example.com/users/Admin@org2.example.com/msp/peer/crypto/
ADDRESS=peer1.org2.example.com:7051ithub.com/hyperledger/fabric/peer# CC
LOCALMSPID="Org2MSP"pt/gopath/src/github.com/hyperledger/fabric/peer# CC
Organizations/org2.example.com/peers/peer1.org2.example.com/tls/ca.crt/y
.org2.example.com:7051/gopath/src/github.com/hyperledger/fabric/peer# ec
peer1.org2.example.com:7051
code query -C home -n HomeACL -c '{"Args":["query","DevA"]}'ric/peer# pe
2018-03-21 16:27:07.843 UTC [msp] GetLocalMSP -> DEBU 001 Returning exis
2018-03-21 16:27:07.843 UTC [msp] GetDefaultSigningIdentity -> DEBU 002
y
2018-03-21 16:27:07.843 UTC [chaincodeCmd] checkChaincodeCmdParams -> IN
2018-03-21 16:27:07.843 UTC [chaincodeCmd] checkChaincodeCmdParams -> IN
2018-03-21 16:27:07.843 UTC [msp/identity] Sign -> DEBU 005 Sign: plaint
0510...4C1A0D0A0571756572790A0444657641
2018-03-21 16:27:07.844 UTC [msp/identity] Sign -> DEBU 006 Sign: digest
90F623DE2012DA22029CDB0CA5C928B9
Query Result: 7
2018-03-21 16:27:24.938 UTC [main] main -> INFO 007 Exiting.....
root@4a5853cfeac7:/opt/gopath/src/github.com/hyperledger/fabric/peer# ▮
```

Fig. 7 Inquiring device authentication

even if transactions of other devices are conducted in the block chain network that has Device C and the device policy is already distributed.

5 Conclusion

Smart Home connects home appliances through wired or wireless home networks, providing the users with safe and secure lives. However, Smart Home environment is resource-limited and has increasing security threats including unauthenticated access, malicious code, and hacking when it is connected to many home appliances. The security threats often lead to personal information and data leakage. In addition, when an error occurs in the server, the service is halted, which is one of the biggest problems of the pre-existing centralized authentication system. Also, in a server-based platform system that searches for devices based on ID, there is a problem that when the components do not work, the entire system is stopped.

In this paper, we proposed a security service framework for user authentication and mutual authentication between devices to securely control the services provided in the smart home environment. As for Smart Home authentication Service, USIM information can be used to perform user identification and authentication process for the device. Also, digital signatures are used to authenticate the device trying to access with the use of the Home token method. The system proposed in the paper enables mutual device authentication through the private key corresponding to the public key. In the proposed system, it is possible to perform device mutual authentication by registering the authentication information in a block chain and distributing it. In conclusion, we expect the proposed framework for the Smart Home device mutual authentication will spread in the not-so-distant future, encouraging further research and development of the security of Smart Home.

Acknowledgements This research was supported by Basic Science Research Program through the National Research Foundation of Korea (NRF) funded by the Ministry of Education (NRF-2017R1D1A3B03036130).

References

1. Ministry of Trade, Industry and Energy, Korea Institute of Design Promotion, Smart Home Trends in industrial environments and related technologies (2016)
2. Strategy Analytics, US leads smart home adoption globally (2014)
3. Internet of Things Forum, Smarthome international standards guidelines for product development, Dec 2016
4. TTA, 2017 Top 10 ICT standardization issues, Dec 2016
5. KISA, Home/Home Appliance IoT Security Guide, July 2017
6. KISA, Device authentication service application for information and communication devices, Sept 2011
7. Eom, J.Y.: Home IoT/ security technology for home appliances. Inf. Commun. Mag. (2017)

8. KISA, Home network construction authentication security guide, July 2017
9. ETRI, A study on the effect of policy changes to communication market, Korea Communications Commission, Nov 2009
10. Gartner, Top 10 Strategic Technology Trends for 2018, Oct 2017
11. Bitcoin, Bitcoin: A Peer-to-Peer Electronic Cash System, pp. 1–9 (2008)
12. Lee, D.Y., et al.: Block chain core technology and domestic and foreign trends. Commun. Korean Inst. Inf. Sci. Eng. **35**(6), 22–28 (2017)
13. Antonopoulos, A.M.: Mastering Bitcoin: Programming the Open Block Chain. O'Reilly Media (2017)
14. Szabo, N.: Smart contracts: formalizing and securing relationships on public networks, First Monday **2**(9) (1997)

A Study of Groupthink in Online Community

Nhu-Quynh Phan, Seok-Hee Lee, Jin Won Jang and Gwang Yong Gim

Abstract Groupthink phenomenon is getting a lot of consideration and many authors have studied about it in the historical research. Base on the development of information technology, people can study, entertain and work together in online communities on the Internet. Consequently, this paper studies about Groupthink phenomenon that can affect the decision-making quality of online communities in Korea. This empirical study towards the respondents was in Korea who currently participate online communities. The total number of surveys collected 285 samples, after removing the invalid samples, and 249 samples were used to conduct the analysis. Our analysis shows that organizational structural faults does not affect overestimation of the group, cohesion of a group also does not affect closed-mindedness and overestimation of the group also does not affect symptoms of defective decision-making. The results achieved in this study are potentially essential for future researches for online communities.

Keywords Groupthink · Online community · Decision quality

1 Introduction

In social psychology, group decision-making sciences and organizational theory, groupthink is a common utilized theory. Moreover, Groupthink is a term created by Irving [1]—a social psychologist. It happens as a group make inaccurate decisions

N.-Q. Phan (✉) · J. W. Jang · G. Y. Gim
Business Administration, Soongsil University, Seoul, Korea
e-mail: phannhuquynh@hotmail.com

J. W. Jang
e-mail: jiwon.jang365@gmail.com

G. Y. Gim
e-mail: gygim@ssu.ac.kr

S.-H. Lee
Information Technology Policy Management, Seoul, Korea
e-mail: chris@lghitachi.co.kr

© Springer International Publishing AG, part of Springer Nature 2019   149
R. Lee (ed.), *Computational Science/Intelligence & Applied Informatics*, Studies
in Computational Intelligence 787, https://doi.org/10.1007/978-3-319-96806-3_11

because the group pressures can cause a decline of reality testing, moral judgment and mental efficiency. Groupthink influent groups by driving them to take foolish actions which dehumanize other groups and ignore alternatives. According to Hassan [2], Group members aim to minimize collision and reach a unity decision without critical evaluation of viewpoints or alternative ideas, and by isolating themselves with external influences.

An online community also considered is a community's virtual form of paralleled with developments of Internet revolution in the worldwide [3, 4]. Online communities are open collectives of members who are identifiable or not necessarily known and they share common interests, and the communities attend to both their collective welfare and both their individual [5].

Samer Faraj et al. proposed that an online community is considered as a virtual organization in which knowledge collaboration occur in unparalleled scope and scale, in ways not previously theorized. For instance, collaboration may occur among people not known together, who share various interests without dialogue [6].

It is assumed that an online community is considered as a virtual community whose members interplay with each other primarily by the Internet. Those who wish to be a part of online communities generally have to become a member via a explicit site and essentially need an internet connection. Online communities can also act as an information system in which members can comment, comment on discussions, give opinion or collaborate. Online communities have become a very popular way for people to interact, who have either known each other in real life or met online. The most common forms people communicate through are chat rooms, forums, e-mail lists or discussion boards. Most people rely on social networking sites to communicate with one another but there are many other examples of online communities. People also join online communities through video games, blogs and virtual worlds.

In the past, Groupthink phenomenon has been studied for the offline working groups, but nowadays, with the high speed of information technology development, people usually study and work with the online groups via the Internet. Therefore, this becomes our motivation to do research the study about the effects of Groupthink for online groups, especially for Korea online communities.

This study try to explore and survey Groupthink phenomenon has affected how the quality of decision making for the online group to find out the causes affecting the quality of group decisions, thereby, create the basis for future studies continue to develop research on the phenomenon of Groupthink in open collaboration platform. Methodological and theoretical issues for future groupthink research are discussed and identified.

2   Literature Review

2.1   Historical Research of Groupthink

2.1.1   Janis' Groupthink Theory

Groupthink will occur if there is the pressure about conformity and cohesiveness, and conformity affect analysis and decision-making of the group, thus leading to a poor decision-making. There is a survival of the victims of Groupthink such as creativity, independent thinking will be lost. Based on Janis's analysis in historical decision-making activities of groups or government policy-making committees groupthink was built and supported by associated content analysis of various political-military successes and fiascos adopting his groupthink model. The political-military fiascos that Janis investigated include: the failure prepared for the attack on Pearl Harbour (1941), Nazi Germany's decision to overrun the Soviet Union (1941), the immature decision on the aggression Bay of Pigs in Cuba (1961), the stalemate in the North Korean War (1950), the escalation in the Vietnam War (1975) and Watergate Cover-up scandal (1972). Janis also studied cases that Groupthink was avoided, for instance, the formulation in the Marshall Plan (1948) and the Cuban Missile Crisis (1962) [1, 7].

Janis's groupthink model (1982) as Fig. 1, consists of three major components: Consequences resulting from symptoms, Antecedent conditions, and Observable Symptoms. In brief, if all antecedent or partial conditions are presented then concurrence-seeking tendency will increase and finally induce groupthink. According to Janis's groupthink model, the potential fatal flaw decision-making often occur with the co-occurrence of high or moderate group cohesiveness and some ante-cedent conditions in group. Janis divided such antecedent conditions into two categories: provocative situational context and structural faults of the organization. In Fig. 1, these two conditions if combined with group cohesiveness can induce Groupthink. Groupthink with eight common symptoms will cause defective decision-making. Defective decision-making is detected by seven observable symptoms. Eventually, Groupthink will reduce the possibility of successful results. In Janis's groupthink model also shows that the antecedent factors are divided into three main groups: structure faults of the organization, cohesiveness and provocative situational contexts. In Fig. 1, the tendency of groupthink seems to increase with the existing of a group with moderate or high cohesiveness interacting with other structural errors of the organization. Janis propose that groupthink tendency will increase when the group is overcoming high stress of expecting damage from external threats. The group seems to favor the leader's solution with no looking for alternatives despite of leader's solution is not a good decision. Janis suggests that cohesiveness is concerned to be the most important condition for groupthink. Indeed, Janis suggests that cohesiveness is concerned to be the most important condition for groupthink. Indeed, Cohesiveness connects members to team up and make them remain in a group. Janis indicates that the power of cohesiveness is caused by interpersonal attraction among members, group commitment to the group's pride and task or mission [1, 7].

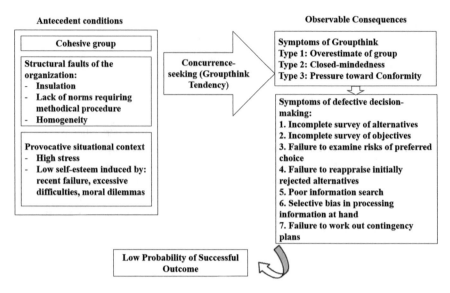

Fig. 1 Janis' Groupthink model (1982) [7]

In this model, the presence of more antecedent conditions is related to the chance of defective decision making by groupthink. The symptoms of groupthink or observables consequences are also indicated in this model. The probability of exhibiting symptoms on defective decision-making will high when a group performs almost of symptoms of groupthink. Poor decision-making often reduces the chance of making an effective decision.

2.2 Groupthink in Online Communities

2.2.1 Definition of Online Communities

The first and presumably the most cited definition of online community was born in 1993 by Howard Rheingold, who described them as rising social gatherings from the network when enough people make public discussions long enough, with enough human senses [4]. Jenny Preece (2000) compared with online communities from the administrator's aspect, impressing that developing them creates a practical activity and that the determination of online communities is necessary to conduct the practice. According to Preece, an online community usually consists of people interacting socially and sharing a common purpose, and includes policies to guide members interacting with each other, is also of the computer system to make people feel connected [8]. In social scientific theories, definitions of community have highlighted the importance of meanings and experiences within a community over the role of appearances and structures. In the notion of symbolic communities introduced by Cohen (1985), a community is constructed symbolically through shared meanings, culture, norms and exists in the minds of its members [9]. Later, Blanchard and Markus (2002, 2004) defined shared emotional connection and a feeling of belonging in a group, a sense of community, as another distinctive feature of online communities [10, 11].

Although the large amount of study on the topic, the phrase "online communities" has been the subject of controversy, as the question whether communities not has been addressed or can exist online by a number of scholars [12, 13]. This is partly explained by the individual associations of the word 'community' such as 'warm and blurry' [8] but also by the distinction between online and geographic communities, particularly is when it comes to intimacy and sharing history among community members [12–14]. These research has found that community feelings or feelings can be fully experienced online. However, not all sites can be labeled as online communities, nor will they eventually become [10]. In practice, it can be seen that the lack of activity as well as lack of user input is the main reason most cited for causing the failure of the online community [15]. Therefore, scholars have acknowledged that user participation is a very important.

According to Hercheui [16], highlighted emphasis has been assigned on the newness of the phenomenon and research on online communities has so far been described rather than theory-driven. There is still a lack of consistency in this area because of the problem that a wide range of community types differ in purpose, structure, and user base compared in the online communities [17]. The main challenge of the study is to present a perspective at a certain point, but not the nature of the topic is constantly evolving [18]. Such snapshots do not adequately and accurately reflect the dynamics of online communities. Currently, online community research is only in the exploration phase, and online communities are growing and dynamic, and more research is needed to improve the generality of the results [17].

2.2.2   Types of Online Community

While the number of online communities keep going to increase and millions of people join in them, researchers have aimed to classify communities to better study them. They have generally characterized between communities based on the need they execute. Hagel and Armstrong [19] proposed that online communities accomplish various needs at different time in a nonexclusive way. A community can be interested in relationship, imagive or transaction. The community types are categories by the specific location, be used by the near users, in the employee class specific, form to be or by the people in age, gender, lifestype or individual objects, according to the specific such as shopping, financial activities or playing game [20]. Lazar and Preece proposed that existing online communities can also be divided into four dimensions such as Table 1 [21].

The attributes of a community involve its goals, type of interaction, topic of interest, type of activity, size, level of anonymity, level of support, type of conventions,

Table 1 Different online communities based on four dimensions [21]

By attributes	By relation to physical communities
Goal, interests [20]	Based on (frequent face-to-face)
Family and lifestyle	City
Work	Government
Play	Education
Spirituality and health	Somewhat (periodic face-to-face)
Politics	Online scholarly community
Business transactions	Hobbies
Education	Not related (no face-to-face)
Intense interaction, emotional ties	Anonymity—role playing
Shared activities	Health
Support	Victims of crime
Conventions, language, protocols	
Anonymity levels	
Sources of revenue	
By supporting software	**By boundedness**
Listservs	Tightly
Newsgroup	Organization intranet
IRC	Loose
MUD	
Web-based bulletin	
Team rooms	

protocols and language among others by Lazar and Preece [21]. As for the relative to physical communities, online communities may require continual, cyclic, or no face-to-face interactions. Online communities use such technologies and software applications as email lists, bulletin boards, newsgroups, meeting rooms and Internet-relay chat and can be loosely or tightly bounded to an organization. Lazar and Preece [21] proposed that the label one uses to indicate a community may alter as each community performs one or more of the features in each of these four dimensions [21]. In instance, krebsgemeischaft.de, a community of cancer patients, considered as high-interactivity community or as a support because it gives support for cancer patients and their families (purpose) and encourages extreme interactions among members (interaction). It also may be considered as privacy-oriented because is supports mechanisms to ensure identity (level of anonymity) and members' privacy, or as an online community with periodic face-to-face interaction or as one with discussion forum (supporting software) because it aims to patients combine with a specific hospital in Germany (physical community) [22].

Others build on the categorizing provided by Lazaar and Preece [23]. source of revenue was added by Leimeister and Krcmar (2004), such as membership revenue, usage-based revenue or subscription-based revenue. other supporting software possibilities such as mailing lists, discussion forums, usenet news, chats, e-groups and immersive graphic environments were added by Preece and Maloney-Krichmar [24]. Areas of interest such as spirituality, work, health, education and politics as goals of online communities were added by Kim [20]. Online community transactions are growing fast, transactions are seen as a target for some online communities [25, 26].

In an investigation of 50 online communities, five genres of communities were identified by Hummel and Lechner [27]. These genres are games, three other mixed genres and interest or knowledge, also oriented to transactions, business-to-consumer (interest, commerce, and transactions), business-to-business (knowledge and transactions), and consumers-to-consumers (interest, transaction and trade). These genres are based on four dimensions that indicate a community, namely, a defined group of actors, sense of place, interaction and bonding. Each of these dimensions exists in each community in the form of management and feature activities. In instance, if a community has a precise content focus and access rules and clear entry. It will have a clearly defined group of actors. Hummel and Lechner [27] work is suitable for our review because they disclosed the basis to translate four dimensions of online communities into physical features (technology and management) that can be executed in online communities [27]. Hummel and Lechner's [27] work can be used to prescribe the highlight and the implementation importance of particular success factors in each of the various community genres

Between 2005 and 2007, an explosion of a new kind of online communities known as social networking sites emerged. These social networking sites are online communities whose only purpose is the maintenance and creation of friendships or social relationships. Because of the development in this new type of online communities, compared with the limited development of traditional communities of interest, it liked that this community type would become the most pervasive. Members of these social networking sites apply Web 2.0 technologies and multimedia such as social bookmarking video sharing and photo to build their profiles and bring themselves to other members [28]. Members invite other members and provide detailed electronic profiles to become their electronic acquaintances (or help others create their profiles, like YahooGraffiti.com). The highlight on these social networking sites primally is including them in networks of connected friends and on meeting people. The most significant examples of social networking sites are Facebook.com and MySpace.com. Founders of these communities assumed that they had found the key for motivate member participation which would lead to successful communities. Otherwise, safety and privacy concerns of members with limited return of investments decelerated the development of these communities. This decreasing is making developers of social networking sites centralize their efforts on developing vertical social networks of members with similar personal interests. Vertical social networks often function in the same way as traditional online communities, such as football fans, pet lovers, interactive video creators for exchanging information with each other [29, 30]. Indeed, in essence social networking sites are online communities that take benefit of the improved and new social computing technology for multimedia and interaction information exchange. In a comprehensive overview of social computing Parameswaran and Whinston [28] proposed that Web 2.0, social computing and online communities are different terms that all refer to those services and applications which "facilitate collective action and social interaction with rich exchange of multimedia information" [28].

Finally, according to Sanna Malinen [31], Online communities types are involved as a specific hobby or health, communities of practice intended for professionals or learning, communities of transaction or enterprise communities, social network sites, creative communities, wikis including question– answering sites and open-source software development.

2.2.3 Online Groupthink Research

There is considerable evidence confirming the symptoms which Janis uses to deter-mine Groupthink—or as sociologists have considered—"Pluralistic Ignorance", are those in the online community as in the traditional group structure [32]. For the peer numbers, such as the chat rooms, be defined with the social community—all people are unrecognized missions insistency with the online group. This is an invalid three keys cannot be missing to guessing Groupthink that is the following tasks of the following users, remove the inconsistency count and not recommended any other comments of each individual [33]. However, the confirm of frequency of expressing their actual beliefs and member preferences is difficult, if possible, to gauge because there is a condition of anonymity in these cases. Indeed, the perceived invulnerability attributed to Groupthink may be purely a virtue of participant anonymity in these cases.

Another example, a group may be working towards problem resolution or devel-oping new projects for issues in their clients' projects or even current projects. It becomes that a efficient and solid decision-making process is necessary for business success. Information may be stored in wiki software to help weight their options bet-ter and employees may need to hold multiple meetings. The idea of building a wiki paltform for business collaboration, though requiring a lot of management culture changes [34], but is generally considered beneficial [35, 36]. In fact, groups are easy to form in any community. Smith [37] provides a good definition of a group primally based on a couple of criteria.

A group is defined as the largest set of two or more members who are specified by a network of suitable communications, a shared sense of collective identification and one or more shared goal dispositions with relative normative strength.

The above statement indicates that groups not only are simple to form but also they can vary in size. Levine and Moreland [38] given Another definition. They stated that a group is several people that "interact on a regular basis, have affective ties with one another, share a common frame of reference and are behaviorally independent" [38].

As mentioned above, it becomes clearly that social media are bound to have groups within them whether an online guild of members in a virtual world or a collaborative project for a company. Anywhere one can join groups, group phenomena is bound to exist and this rule can applies to social media. However, people can change the virtual environment created by social communication software, no same as the commassers in the world, where all restricted user is allowed to verify environment. In fact, the software is entirely manufactured. Indeed, a question comes to mind is and could we change the software by a way that we can change group interactions, and as such protect groups from dangerous group phenomena such as groupthink?

Michail Tsikerdekis [39] proposed that there is a risk with Groupthink behavior in online groups or group decision support systems (GDSS), when almost all potential alternatives are not considered for problem resolution. It becomes an actuality when individuals simply hesitate to propose their own solutions to a problem and adapt to the majority opinion. Base on a survey of random participants from the English-language Wikipedia community, he founded the effects of anonymity on the likelihood of adapting to group opinion. Moreover, a qualitative study was carried out concurrently in his study. He sought to examine existing online communities that collaborate under several anonymity states. Michail Tsikerdekis assumed that Groupthink can be less likely to occur online due to the likelihood of individuals disclosing potential alternatives instead of simply adapting to group is higher and the occurrence of alternatives is vital for preventing groupthink. The importance of having various alternatives to solve problems in a group is to be an major part of problem solving in groups. It is unclear for a group when individuals are arguable from disclosing their opinions. In additional, his findings clearly indicate that even if effect is not very recognized, when getting anonymity rises, the ability for the selected options will be grow up. So he suggested that this is a very important relationship that software developers should not ignore.

There is a case study called Alternate Reality Games (ARGs), or World Without Oil created in an online environment, Nassim Jafari Naimi and Eric M. Meyers has proposed that ARG is a collective environment that consideration is given to features that go against the goal of collective intelligence, making the environment more vulnerable to the group [40].

A study by Lawrence Leung [41] suggested that the impact of participating in role simulation in online games tended to be divided into five questions groups. With the result of the analysis, it is said that participants have gained a reduction of Groupthink tendency after the online role play simulations exercise. This result also assumes that role simulations in online games have a positive effect in reducing Groupthink trends.

In traditional group interaction, where the normative community concept precedes the inclusion of online social networks, research shows that Groupthink contributes to negative organizational outcomes. In certain cases, like the Challenger incident, these consequences have proven to be quite severe—including the tragic loss of human life and financial ruin [42]. The findings in this paper also reveal that many of the same elements present within traditional group interaction are, in varying degrees, also present on computer-mediated communities like Facebook. Thus, understanding how and to what degree Groupthink will emerge in the online landscape is a research topic of tremendous importance as decision-makers are steadily incorporating social networking into their organizational infrastructure In addition, the expanding sphere of influence such networks have on organizational group structures creates fascinating new opportunities for qualitative and quantitative research in this emerging field of academic study. Further studies may reveal more specifically how the Groupthink symptoms translate into computer-mediated environments and offer modernized solutions relevant to identifying and preventing these conditions.

3   Research Model and Hypothesis

Based on Groupthink model of by Janis proposed in 1982, we have conducted empirical studies to verify Groupthink phenomenon in Online communities. The research model (Fig. 2) as follows:

Cohesion has been defined as "group members inclination to forge social bonds, resulting in members sticking together and remaining united" [43]. It is one of most widely studied variables and the oldest in the group dynamics literature [44, 45] and is fundamental to the fabric of group. Cohension is defined as the desire of the members to form a group [46] or the way in which individual members of an affiliated group work together [47].

Organizational structure refers to the method in which an organization organizes personnel and work so that work is carried out and that its objectives can be fulfilled [48]. The structural faults are indicated such as insulation of the group, lack of norms requiring methodical procedures, lack of tradition of impartial leadership, homogeneity of members' social ideology and background.

Due to these structural failures, the group tends to become easy prey to what Janis calls "provocative situational contexts" [49]. Provocative situational contexts are made as up the high pressure of the threat of the external, excessive difficulties on current decision-making tasks and moral dilemma, or low self-esteem temporarily induced by the group's perception of recent failures [7].

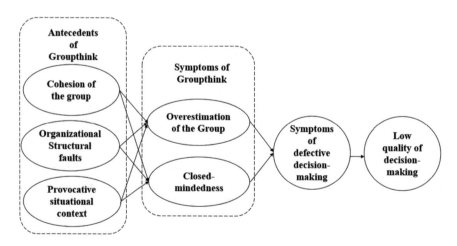

Fig. 2 Research model

Closed-mindedness includes Collective Rationalization and Out-group Stereo-types. Collective Rationalization. In finalizing any decision process, it is natural and normal to downplay the drawbacks of a course. In a group, the problem arises when they see no fault to their plans, even if there is significant evidence as to the folly of their chosen course of action. Moreover, any legitimate objections that may exist are totally overshadowed by the perceived negative reaction that anyone voicing objec-tion would receive. Out-group Stereotypes. Many of us have known this symptom as we were growing up. We and our group of friends thought that we were "cool" and made fun of the "not cool" cliques. Primarily, we formed stereotypes against individuals who were not member of the group. This symptom continues where we stereotype our views on our enemy leaders as incompetent or weak with adults

Overestimation of the group includes the overestimation of the potential success of the solution or the abilities of the group and belief in the inherent morality of the group [2].

Groupthink is a group bias [7]. Both group bias and group error share similar meanings. A group decision bias refers to a deviation between the normative decision-making and group decision-making behavior. The cause of error may not be a bias but it is responsible. Duffy [50] found that cognitive social errors and organizational are the three sources of group error.

Duffy [50] defined Groupthink as a kind of social error. In fact, Groupthink is the worst bias in terms of damages in crisis decision-making among the group errors mentioned above.

Therefore, the following hypotheses are proposed:

- H1: Cohesion of the group will have a positive effect on Overestimation of the group in Online communites
- H2: Cohesion of the group will have a positive effect on Closed-mindedness in Online communites
- H3: Organizational structural faults will have a positive effect on Overestimation of the group in Online communites
- H4: Organizational structural faults will have a positive effect on Closed-mindedness in Online communites
- H5: Provocative situational context will have a positive effect on Overestimation of the group in Online communites
- H6: Provocative situational context will have a positive effect on Closed-mindedness in Online communites
- H7: Overestimation of the group will have a positive effect on Symptoms of defec-tive decision-making in Online communites
- H8: Closed-mindedness will have a positive effect on Symptoms of defective decision-making in Online communites
- H9: Symptoms of defective decision-making will have a positive effect on Low Quality of Decision-making in Online communites

4   Research Method and Analysis Results

A survey designed by basing on the research conceptual was developed for this study. The hypotheses will be tested follow response from the direct survey. First, we employed Confirmatory Factor Analysis (CFA) to estimate the validity of the measurement for the model then use Structural Equation Modeling (SEM) to test the proposed model, so that we evaluated the causal structure of the model. AMOS 20 was used to analyze the structure model and measurement model.

4.1   Demographic

A total number of 249 responses were utilized in the analysis. The demographic characteristics and general statistical characteristics are following in Table 2.

4.2   Reliabilities Analysis

The reliability of the questionnaire scale was tested using Cronbach's alpha for total of forty measurement items, which divided into 7 factors. The Cronbach's alpha for scales in the model range from 0.739 to 0.946. hese results suggest that measurements for the scale are reliable. Specific indicators for reliability analysis for each factor are presented in Table 3.

4.3   Factor Analysis

Reliability and Validity were examined by computing Cronbach's alpha coefficient for the construct. Convergent validity and discriminant of the scales was initially examined using extraction method is principal axis factor analysis with Varimax. The items loaded significantly on their hypothesis factors. The factor analysis for each construct is presented detailly in Table 4.

4.4   Confirmatory Factor Analysis

This study used CFA (Confirmatory Factor Analysis) to test scale based on the EFA (Exploratory Factor Analysis) results.

Table 2 The descriptive statistics

Item	Frequency analysis	
	Frequency	Percent (%)
Gender		
Male	190	76.3
Female	59	23.7
Total	249	100.0
Age		
20–29	92	36.9
30–39	48	19.3
40–49	88	35.4
50–59	20	8.0
Over 60	1	0.4
Total	249	100.0
Education		
Under graduated student	77	30.9
Graduated student	63	25.3
Master/Ph.D. student	58	23.3
Master/Ph.D.	51	20.5
Total	249	100.0
Working experience		
Under 1 year	88	35.3
1–5 years	26	10.4
5–10 years	34	13.7
10–15 years	32	12.9
Over 15 years	69	27.7
Total	249	100.0
Working position		
Staff	44	17.7
Manager	60	24.1
Senior manager	33	13.3
Other	112	44.9
Total	249	100.0
Income		
Under 15 million won/year	87	34.9
15–25 million won/year	9	3.6
25–35 million won/year	4	1.6
35–50 million won/year	24	9.6
50–70 million won/year	60	24.1
Over 70 million won/year	65	26.1
Total	249	100.0

Table 3 The Cronbach's alpha

Factor	Cronbach's alpha	Items
Cohesion of the group	0.905	5
Organizational structural faults	0.739	4
Provocative situational context	0.873	5
Overestimate of the group	0.903	6
Closed-mindedness of the group	0.811	5
Symptoms of defective decision-making	0.913	7
Low quality of group decision-making	0.946	6

Table 5 is the fit of the confirmatory factor analysis conducted to measure variables after modification factor analysis conducted. Construct reliability (CR) and average variance extracted (AVE) are the components for convergent validity.

Factor analysis confirmed the results of the analysis based on the results came out, and then subjected to a reliability analysis as shown in Table 6 Concept of reliability (CR: Composite Reliability) were satisfied with the level of more than 0.7 and the average variance extracted (AVE: Average variance extracted) were satisfied with the level of more than 0.5 (Fornell and Larcker 1981), the research model of this study was to secure the concept of reliability.

The correlation matrix, with correlations among constructs and the square root of AVE on the diagonal is shown in Table 7 Adequate reliability, convergent validity, and discriminant validity were demonstrated in the measurement model

4.5 Hypothesis Testing

In this study, we confirmed the structural equation modeling using AMOS to test hypotheses, verify relationships with key variables are shown in Table 8 and Fig. 3.

5 Conclusion

This study explores the relationship between symptoms of Groupthink, antecedents of Groupthink, low quality of decision-making and symptoms of defective decision-making in online communities. Our study demonstrates the validity of the theory model of Janis [7] but the result also found a surprising result that Cohesion of the group not effect on Closed-mindedness in Online communities. This indicated that supportive Cohesion of the group is not the reason to occur Closed-mindedness that

Table 4 The exploratory factor analysis (EFA)

Component	1	2	3	4	5	6	7
CO1				0.773			
CO2				0.798			
CO3				0.803			
CO4				0.745			
CO5				0.836			
ORG1							0.724
ORG2							0.628
ORG3							0.562
ORG4							0.688
PRO1					0.735		
PRO2					0.693		
PRO3					0.739		
PRO4					0.752		
PRO5					0.727		
OVER1			0.792				
OVER2			0.862				
OVER3			0.760				
OVER4			0.785				
OVER5			0.729				
OVER6			0.641				
CLO1						0.686	
CLO2						0.780	
CLO3						0.634	
CLO4						0.704	
CLO5						0.642	
SYM1	0.675						
SYM2	0.578						
SYM3	0.762						
SYM4	0.707						
SYM5	0.690						
SYM6	0.734						
SYM7	0.671						
QUA1		0.770					
QUA2		0.689					
QUA3		0.759					
QUA4		0.745					
QUA5		0.761					
QUA6		0.733					

Table 5 Model fit indices

Model fit indices	CMIN/DF	CFI	GFI	AGFI	RMSEA	TLI
Recommended value	<3	>0.8	>0.7	>0.7	<0.08	>0.8
Obtained	2.178	0.901	0.790	0.757	0.069	0.892

Table 6 Reliability Analysis of confirmatory factor analysis

Item	CO	ORG	PRO	OVER	CLO	SYM	QUA
CR	0.906	0.789	0.879	0.899	0.832	0.914	0.947
AVE	0.660	0.557	0.647	0.642	0.558	0.604	0.748

Table 7 Discriminant validity

	CO	ORG	PRO	PRE	CLO	SYM	QUA
CO	**0.604**						
ORG	0.762	**0.748**					
PRO	−0.408	−0.461	**0.642**				
PRE	−0.349	−0.436	0.609	**0.660**			
CLO	0.625	0.644	−0.134	−0.298	**0.647**		
SYM	0.679	0.581	−0.176	−0.143	0.552	**0.558**	
QUA	0.624	0.570	−0.308	−0.327	0.606	0.549	**0.557**

Table 8 Hypothesis test

	Path coefficient	S.E.	C.R	P	H-test
OVER <— CO	0.591	0.060	8.129	***	Supported
CLO <— CO	0.076	0.047	0.883	0.225	**Rejected**
OVER <— ORG	−0.213	0.100	3.034	0.016	**Rejected**
CLO <— ORG	0.360	0.107	4.533	***	Supported
OVER <— PRO	0.152	0.081	1.690	0.055	Supported
CLO <— PRO	0.416	0.080	5.180	***	Supported
SYM <— OVER	−0.328	0.050	−0.087	***	**Rejected**
SYM <— CLO	0.681	0.088	11.628	***	Supported
QUA <— SYM	0.778	0.069	12.352	***	Supported

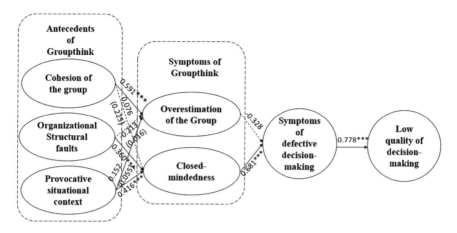

Fig. 3 Researching results

Janis and the other researchers have asserted. Moreover, it is important to find that the negative effects of directive Groupthink behavior found in this study are not related to cohesiveness. Most likely, cohesiveness of the online group members contributes in positive ways to team performance and individual, such as by the effectiveness of the group decision-making and increasing productivity.

Furthermore, this study also found that the organizational structural faults does not affect on the overestimation of the group and excessive the overestimation of the group does not affect on the symptoms of defective decision-making in online communities.

The explanation for these finding may be concerned to the characteristics of interaction and communication of the online communities. Initially, the offline groups in collectivist cultures are able to be perceived by its members as a weak social identity, such as this given the group composition of people with different societal background and status, age, and education. But in offline groups, traditional values of sacrificing for the sake of a group are more applicable compared to other online groups. Moreover, high cohesiveness is not inherent to the group in the collectivist societies until crucial conditions for its development are suitable. In turn, online groups cohesiveness might be high, due to the comparatively high importance of work for individuals as the job can suitable for a wide set of collaborative needs. In such cases group members in online groups might develop positive attitudes towards self-management and teamwork, and move toward a collectivist orientation. In any case, results showed that low or high team cohesiveness of work groups should not be considered as an inherent part of Groupthink. Therefore, its development requires additional attention of some other factors. Finally, according to this results, development of group cohesiveness in online communities should be attention in future research.

5.1 Limitation

In this study, we only surveyed within a small object with only a brief survey of the people are living, studying and working in Seoul, South Korea that we still can't deploy a wide across many other countries.

Based on a theoretical model of the Groupthink phenomenon suggested by Janis [1], we just conducted an empirically verifiable to consider this model is suitable for online communities or not without finding additional variables that may have an impact on the quality of group decision- making in online communites.

5.2 Future Research

We will expand the scope of survey in many different countries with more number of surveys to improve the quality of data collected.

In addition to the factors proposed by Janis, we will expand the search for other factors that influence the decision-making process for the online groups as well as the quality of the decisions of online communities.

References

1. Irving, L.J.: Victims of Groupthink: A Psychological Study of Foreign-Policy Decisions and Fiascoes. Houghton Mifflin Company (1972)
2. Hassan, G.: Groupthink principles and fundamentals in organizations. Interdisc. J. Contemp. Res. Business, **5**(8) (2013)
3. Don, T., Anthony, D.W.: Wikinomics: How Mass Collaboration Changes Everything. Portfolio, New York (2006)
4. Howard, R.: The Virtual Community: Homesteading on the Electronic Frontier. MIT Press, Cambridge, MA (2000)
5. Lee, S., Manuel, A.: Online communities. Bidogli, H. (ed.) Handbook of Computer Networks, vol. 3. Wiley, New York (2007)
6. Samer, F., Sirkka, L.J., Ann, M.: Knowledge collaboration in online communities. Organ. Sci. **22**(5), 1224–1239 (2011)
7. Irving, L.J.: A revised and enlarged edition of victims of groupthink: a psychological study of foreign-policy decisions and fiascoes (1972). Psychological Studies of Policy Decisions and Fiascoes (1982)
8. Jenny, P.: Online Communities: Designing Usability and Supporting Sociability. Wiley (2000)
9. Anthony, P.C.: The Sympolic Construction of Community. Ellis Horwood Limited (1985)
10. Blanchard, A.L., Markus, M.L.: Sense of virtual community-maintaining the experience of belonging. In: Paper Presented at the 35th Annual Hawaii International Conference on System Sciences (2002)
11. Blanchard, A.L., Markus, M.L.: The experienced "sense" of a virtual community: characteristics and processes. Database Advanc. Informat. Syst. (2004)
12. Wittel, A.: Toward a networking sociality. Theor. Cult. Soc. (2001)
13. Miller, V.: Understanding Digital Culture. Sage, London (2011)

14. Brint, S.: Gemeinschaft revisited: a critique and reconstruction of the community concept. Sociol. Theor. (2001)
15. Ling, K., Beenen, K., Ludford, P., Wang, X., Chang, K., Li, X.: Using social psychology to motivate contributions to online communities. J. Comput. Mediat. Commun. (2005)
16. Hercheui, M.D.: A literature review of virtual communities: the relevance of understanding the influence of institutions on online collectives. Informat. Commun. Soc. (2010)
17. Gallagher, S.E.: Savage: cross-cultural analysis in online community research: a literature review. Comput. Human Behav. (2013)
18. Alicia, I., Gondy, L.: A life-cycle perspective on online community success. ACM Comput. Surv. **41**(2) (2009)
19. Hagel, J.I., Armstrong, A.G.: Net Gain: Expanding Markets through Virtual Communities. Harvard Business School Press, Boston, MA (1997)
20. Kim, A.J.: Community Building on the Web. Peachpit Press, Berkeley, CA (2000)
21. Lazar, J., Preece, J.: Classification schema for online communities. In: Hoadley, E., Benbasat, I. (eds.) Proceedings of the Fourth Americas Conference on Information Systems, Baltimore, MD, August. AIS, Atlanta, GA (1998)
22. Jan, M.L., Helmut K.: Engineering virtual communities in healthcare: the case of www.krebs gemeinschaft.de. Elect J. Organization. Virtual. (2003)
23. Jan, M.L., Helmut K.: Success factors of virtual communities from the perspective of members and operators: an empirical study. In: Proceedings of the 37th Annual Hawaii International Conference on System Sciences (2004)
24. Jenny, P., Diane M.K.: Online communities. In: Jacko, J., & Sears, A. (eds.), Handbook of Human–Computer Interaction, Mahwah: NJ: Lawrence Erlbaum Associates Inc. Publishers (2003)
25. Resnick, P., Zeckhauser, R.: Trust among strangers in Internet transactions: Empirical analysis of eBay's reputation system. In The Economics of the Internet and E-Commerce, M. R. Baye, Ed. Elsevier Science, Amsterdam, The Netherlands (2002)
26. Hiltz, S.R., Goldman, R.: Learning Online Together: Research on Asynchronous Learning Network. Lawrence Erlbaum Associates, Mahwah, NJ (2004)
27. Hummel, J., Lechner, U.: Social profiles of virtual communities. In: Proceedings of the 35th Hawaii International Conference on System Sciences (January). IEEE Computer Society Press, Los Alamitos, CA (2002)
28. Parameswaran, M., Whinston, A.B.: Soc. Comput. Overview. Commun. AIS **19**, 762–780 (2007)
29. Ebrahim, E.: Social Networking: Time for a Silver Bullet. Read/writeweb.com (2006)
30. Bajarin, T.: The Future of Social Networking. PC Mag (online) (2007)
31. Sanna, M.: Understanding user participation in online communities: a systematic literature review of empirical studies. Comput. Human Behav. **46**, 228–238 (2015)
32. Westphal, J., Bednar, M.: Pluralistic Ignorance in corporate boards and firms' strategic persistence in response to low firm performance. Adm. Sci. Q. **50**, 262–298 (2005)
33. Thurlow, C., Lengel, L., Tomic, A.: Computer Mediated Communication. Sage Publications, Thousand Oaks (2006)
34. Penny, E.: Managing Wikis in Business. Open University Business School, Milton Keynes MK7 6ZU, United Kingdom (2007)
35. Kussmaul, C., Jack, R.: Wikis for Knowledge Management: Business Cases, Best Practices, Promises, & Pitfalls. Web 20 (2009)
36. Lauren, W.: Blogs & Wikis: technologies for enterprise applications? Gilbane Rep. **12**(10), 1–24 (2005)
37. David, H.S.: A parsimonious definition of "group:" toward conceptual clarity and scientific utility. Sociol. Inquiry **37**(2), 141–168 (1967)
38. Levine, J.M., Moreland, R.L.: Group socialization: theory and research. Eur. Rev. Soc. Psychol. **5**(1), 305–336 (1994)
39. Michail, T.: The Effects of Perceived Anonymity and Anonymity States on Conformity and Groupthink in Online Communities: A Wikipedia Study. J. Am. Soc. Inform. Sci. Technol. **64**(5), 1001–1005 (2013)

40. Nassim, J. N., Eric, M. M.: Collective Intelligence or Group Think? Engaging Participation Patterns in World without Oil. Proceeding CSCW '15 Proceedings of the 18th ACM Conference on Computer Supported Cooperative Work & Social Computing, 1872–1881 (2015)
41. Lawrence, L.: Exploring the effectiveness of online role play simulations to reduce groupthink in crisis management training. A thesis for the degree of Doctor of Education at The University of Hong Kong (2014)
42. Moorhead, G., Ference, R., Neck, C.P.: Group decision fiascoes continue: space shuttle challenger and a revised groupthink framework. Human Relat. (1991)
43. Carron, A.V.: Cohesiveness in sport groups: interpretations and considerations. J. Sport Psychol. (1982)
44. Mullen, B., Copper, C.: The relation between group cohesiveness and performance: an integration. Psychol. Bulletin (1994)
45. Campbell, M.C., Martens, M. L.: Sticking it all together: a critical assessment of the group cohesion performance literature. Int. J. of Manag. Rev. (2009)
46. Banki, S.: Is a good deed constructive regardless of intent? Organization Citizenship Behavior, Motive, and Group Outcomes. Small Group Research (2010)
47. Aoyagi, M.W., Cox, R.H., McGuire, R.T.: Organizational citizenship behavior in sport: Relationships with leadership, team cohesion, and athlete satisfaction. J. Appl. Sport Psychol. (2008)
48. Nedal, M.E., Ahmed, E., Okasha, A., Abdelghaly, A.: Defining and solving the organizational structure problems to improve the performance of ministry of state for environmental affairs—Egypt. Int. J. Scientif. Res. Publicat. 3(10) (2013)
49. Wexler, M.N.: Expanding the groupthink explanation to the study of contemporary cults. Cultic Stud. J. 12(1), 49–71 (1995)
50. Lor, R.D.: Team decision-making biases: an information-processing perspective. In: Klein, G., Orasanu, J., Calderwood, R., Zsambok, C. Decision Making in Action: Models and Methods, Ablex, Norwood, NJ, pp. 346–359 (1993)
51. Michael, S. B.: Groupthink in Software Engineering". International Journal of Computing and Business Research, 5(1) (2014)

Development of a Physical Security Gateway for Connectivity to Smart Media in a Hyper-Connected Environment

Yong-Kyun Kim, Geon Woo Kim and Seoung-Hyeon Lee

Abstract Due to the rapid technological growth of smart media and the rapid spread of convenience through increased convenience, most individuals have smart media, and many person information is stored on a server connected to smart media. As a result, there is always a risk of leakage of personal information, and there is a great need for a method of protecting the personal data of the server from the invasion of the server. In this paper, we propose a method of blocking the intrusion from the external through unidirectional communication method.

Keywords Unidirectional gateway · Security · TCP · DB replication

1 Introduction

With the growth of smart media, a rapid transition to the IoT world is taking place, including bid data and cloud computing. Hyper-connected refers to the communication between person to person and person to machine belonging to a networked organizational society. Computer networks function as a social network, which forms and connects a global network [1, 2]. This hyper-connection scenario has a profound impact on the relationship between individuals, enterprises, and public resources.

It is based on the connectivity and functionality acceptable to next-generation networks, the ability to support and integrate the Internet of Thins, M2M and Cloud computing. In the IoT era, various services are provided through a connection with servers and media. If a malicious hacker attacks a server that has multiple connections

Y.-K. Kim · G. W. Kim · S.-H. Lee (✉)
Information Security Research Division, Electronics and Telecommunications
Research Institute, Daejeon, Korea
e-mail: duribun2@etri.re.kr

Y.-K. Kim
e-mail: ykkim1@etri.re.kr

G. W. Kim
e-mail: kimgw@etri.re.kr

© Springer International Publishing AG, part of Springer Nature 2019
R. Lee (ed.), *Computational Science/Intelligence & Applied Informatics*, Studies
in Computational Intelligence 787, https://doi.org/10.1007/978-3-319-96806-3_12

to a 1:N, it can easily infect many smart media and spread malicious code quickly [3, 4]. Therefore, the security of the server is required more than ever.

In this paper, we propose a robust TCP communication method to prevent external intrusion by communicating in unidirectional communication to enhance the security of the server. In Sect. 2, explains research trends for blocking external intrusion, and Sect. 3, we propose a design direction and configuration method to guarantee TCP reliability. Section 4 show the implementation results of TCP proxy in the unidirectional gateway. Finally, we conclude in Sect. 5.

2 Related Work

This chapter describes the access control method for personal information database for enhancing security [5–7]. There are server agent methods, gateway methods, and sniffing methods to control access to the personal information database.

The server agent method is the most powerful security method that can control all access routes including a dedicated client which access DB directly by installing agent including access control and logging function on the server itself. However, it causes traffic to the DB server, the performance of the DB server is worried about, and the risk of system shutdown is inherent.

The gateway method is divided into a proxy method and an inline method. Proxy method is a method to change all IPs connected to DB server through DB security server (proxy server). It provides the most powerful access control function and it is advantageous for large system environment because target DBMS can be added. In addition, when a security server (proxy server) fails, redundancy can be configured so that it can be restored without hindrance to online business. The inline method is to configure an inline security system between the target DB server and the client network. It does not need to install or change any agent on the server or client. It is not large scale, DB server is located in one place and the proportion of online business is very high It is advantageous for a non-system. However, when the security server is shut down, there is a risk of interruption of all tasks, and a number of DB security servers are required depending on the size of the target DB server.

The sniffing method is a security method in which packets on the network line are analyzed and logged through the TAP method and the packet mirroring method, thereby placing a greater emphasis on the meaning of post-audit. There is no need to install or configure any agent between the server and the client. It is easy to construct the system without a load on the network, but it is difficult to maintain integrity due to data corruption or damage. Each method is shown in Figs. 1, 2 and 3.

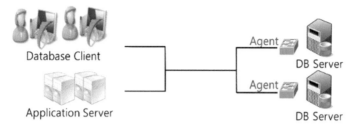

Fig. 1 The agent method

Fig. 2 The gateway method

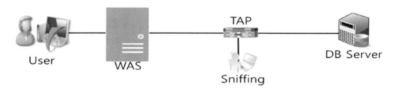

Fig. 3 The sniffing method

3 Design of Unidirectional Security Gateway

In this Chapter, we design a unidirectional security gateway hardware design network interface card (NIC), Database replication method and Retransmission techniques.

3.1 Design of Unidirectional Security Gateway Hardware

A unidirectional security gateway consists of a Tx-only platform, a Rx-only platform, Tx-only and Rx-only network interface card (NIC). Tx/Rx interface card includes bit-level forward error correction (Reed Solomon Encoder/Decoder) and integrity parity functions. And Tx/Rx interface card is developed with a low-power FPGA PCIe interface. Data transmission for 32 channels is possible (Figs. 4 and 5).

Fig. 4 Tx-only/Rx-only platform

Fig. 5 Rx-only NIC

3.2 Design of DB Replication Using CDC in Unidirectional Security Gateway

In general, Information is provided using a database mirroring method to protect the main server's database. Database mirroring is used with the replicate database to provide the availability of the service database. Typically, two single database copies are used on different servers, and clients can always use only one copy of the database. Mirroring involves applying the transaction log of all insert, update, or delete operations performed on the source database to the mirror database.

However, as a number of data increases, the mirroring techniques can cause additional load on the transaction data generated in the system. Therefore, a synchronization technique using data changing tracking of the database is mainly used. CDC (Change Data Capture) method is mainly used for tracking data change. CDC methods include ODBC/JDBC Adapters, SQL Query (using Timestamp to distinguish records to be extracted), Database triggers, and conversion data logging user table, and file comparison (compared with the entire data at the time of the last CDC) and a method of extracting directly from the database log [8, 9].

3.3 Design of the Retransmission Techniques for Non-transport DB Queries

In order to guarantee the minimum reliability of the data, it is created as a data frame suitable for the unidirectional security gateway, so that it is possible to check the omission of data and data contamination [10]. In order to guarantee the minimum reliability of the data, a sequence number is generated in a data frame and sequentially increased. The target system can use this number to determine whether the data is lost or not. The source system also inserts a CRC (Cyclic Redundancy Check) code to check for data corruption.

4 Experimental Result

In this chapter, we have experimentally verified the design contents. We verified the performance of the unidirectional dedicated Tx/Rx-only **NIC** (throughput, multi-channel), verification of communication using unidirectional TCP transmit/receive proxy module, and CDC based DB replication Transmit/receive proxy module.

4.1 Verification for Unidirectional Tx/Rx Dedicated NIC

As shown in Fig. 6, we demonstrated the performance of the 10G Tx/Rx dedicated NIC. We have experimented and measured the overhead including forward error correction and integrity parity check routine.

Figure 7 shows the multi-channel performance results. We confirmed that the data was transmitted and received using 32 logical channels.

Fig. 6 Throughput performance result of 10G Tx/Rx dedicated NIC

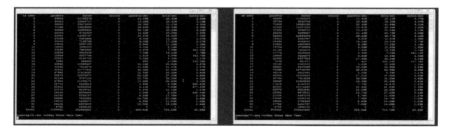

Fig. 7 Multi-channel performance result of 10G Tx/Rx dedicated NIC

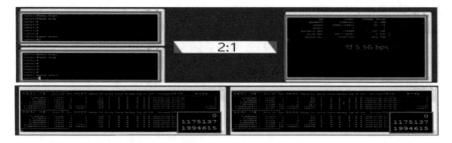

Fig. 8 Performance result of 10G Tx/Rx dedicated NIC

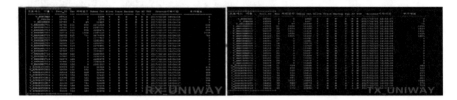

Fig. 9 Data mismatch between Tx and Rx

4.2 Verification for Unidirectional Transmit/Receive TCP Proxy

Section 4.2 presents experimental results for a module that implements a typical bidirectional communication TCP proxy with unidirectional characteristics. Figure 8 shows that our implemented module show 5.5G performance in 2:1 communication, and shows that the data sent from the sender is all transmitted to the receiver.

4.3 Verification CDC-Based DB Replication Transmission Proxy Module

Figure 9 shows the results of the data processing on the TX and RX. TX transmitted a total 7,237, but RX received 4,291.

As shown in Fig. 10 Unreceived data were transmitted through retransmission.

5 Conclusions

A unidirectional security gateway was used as a network connection technology to ensure continuity of work in a physical network separation environment.

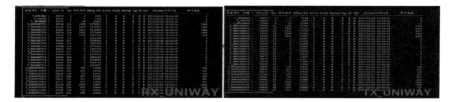

Fig. 10 Data match between Tx and Rx through retransmission

In this paper, we development of a physical security gateway for connectivity to smart media in a hyper-connected environment. the protection of personal information database and provide necessary information to users through database synchronization with service server(destination). Experimental results showed that synchronization of data in the database was successful.

The proposed method can be applied to the data transmission from the safe area to the unstable area where the source data should be protected. Especially, it can be applied to the national backbone network or industrial facilities requiring high security.

Acknowledgments This work was supported by Institute for Information & Communications Technology Promotion (MSIP) grant funded by the Korea government (MSIT) (2018-0-00226).

References

1. Wellman, B.: Physical place and cyberspace: the rise of personalized networking. Int. J. Urban Reg. Res. **25**(2), 227–252 (2001)
2. Freeman, L.: The sociological concept of group: an empirical test of two models. Am. J. Sociol. **98**, 152–166 (1992)
3. Waterfall Security Solutions Ltd. http://www.waterfallsecurity.com
4. OwlComputing Technologies, Inc. http://owlcti.com
5. Eun-Ae, C.H.O., et al.: Database security system for applying sophisticated access control via database firewall server. Comput. Informat. **32**, 1192–1211 (2013)
6. Emil BURTESCU, Database security-attacks and control methods. J. Appl. Quant. Methods **4**, 449–454 (2009)
7. Daniel, C.: Access control in database management systems. Datasem **98**, 153–163 (1998)
8. Change Data Capture. http://en.wikipedia.org/wiki/Change_data_capture
9. Mitchell, J., Eccles et al.: True real-time change data capture with web service database encapsulation. In: IEEE 6th World Congress on Service, pp. 128–131 (2010)
10. Peter Gilbert et al.: The duality between message routing and epidemic data replication. In: Eighth ACM Workshop on Hot Topics in Networks, October (2009)

Design and Implementation of Security Threat Detection and Access Control System for Connected Car

Joongyong Choi, Hyeokchan Kwon, Seokjun Lee, Byungho Chung and Seong-il Jin

Abstract Security vulnerabilities are also increasing as connectivity increases to provide driving stability and convenience for automobiles. In this paper, we design a white list-based access control system to detect and block malicious attempts to access an in-vehicle network through an infotainment device in a connected car environment, and present the implementation results.

Keywords Connected car · Security · In-vehicle infotainment · Infotainment Headunit

1 Introduction

As the environment of automobiles and transportation services develops, vehicles and infrastructures become more complex and more connected. As the connectivity increases, various information can be gathered to improve the stability of the vehicle driving including autonomous driving, and there is an advantage that convenient service can be provided through the infotainment device. However, the importance

J. Choi (✉) · H. Kwon · S. Lee · B. Chung
Information Security Research Division, Electronics and Telecommunications
Research Institute, Daejeon, South Korea
e-mail: choijy725@etri.re.kr

H. Kwon
e-mail: hckwon@etri.re.kr

S. Lee
e-mail: junny@etri.re.kr

B. Chung
e-mail: cbh@etri.re.kr

S. Jin
Chungnam National University, Daejeon, South Korea
e-mail: sijin@cnu.ac.kr

© Springer International Publishing AG, part of Springer Nature 2019
R. Lee (ed.), *Computational Science/Intelligence & Applied Informatics*, Studies
in Computational Intelligence 787, https://doi.org/10.1007/978-3-319-96806-3_13

of related research is increasing because the increase in connectivity can be used as a path of malicious attack through hackers.

In this paper, we design the access control mechanism based on the headunit of the vehicle and the result. The paper is organized as follows. Section 2 describes the connectivity of the vehicle. Section 3 reviews the major weaknesses of the vehicle. In Sect. 4, attack scenario through headunit, access control mechanism proposal, and Sect. 5 summarize the implementation results and future research plan.

2 Automotive Connectivity

Autonomous vehicles can travel independently through various sensors such as Radar, Lidar, and Camera. However, they show various connectivity to enhance the stability of the driving and provide various functions of the driver or passenger [1–3]. Vehicle to Vehicle (V2V), Vehicle to Infrastructure (V2I), Vehicle to Nomadic Device (V2D), Vehicle to Network (V2N) and In-Vehicle Network (IVN) As the environment becomes more difficult, the operating platform of the vehicle platform is becoming more open and standardized.

As the vehicle platform becomes operational and open, the development productivity will increase. However, since the accessibility is improved not only for the manufacturer but also for the general public, researchers and hackers, the vulnerability of the related platform is analyzed and it can not be used easily for the malicious attack.

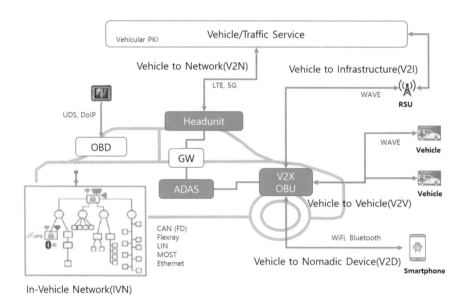

Fig. 1 Automotive connectivity

In-Vehicle Network uses CAN, LIN, Flexray and MOST, and recently Ethernet is also being introduced. In the whole field of the vehicle, important system such as Brake, Transmission, and Engine are composed of CAN. It is especially important because once you can access to the characteristics of CAN, you can freely adjust vehicle such as vehicle steering or brake [4–9] (Fig. 1).

3 Security Vulnerability

The main security weaknesses of the vehicle are electronic control unit (ECU), mobile application, OBD-II port, CAN (Controller Area Network) bus, wireless network, embedded application, CD player and USB port [10].

In Washington University's paper, we classify vehicle threat models into three categories: indirect physical access, short-range wireless access, and long-range wireless. Indirect physical access includes OBD-II, Entertainment (Disc, USB and iPod), Short-range wireless access includes Bluetooth, Remote Keyless Entry, Tire Pressure Monitoring System (TPMS), RFID car keys and Emerging short. Finally, long-range

Table 1 Major security vulnerabilities in vehicles

Vulnerability	Description
ECU	• It can be re-programmed with malicious software
Mobile App'	• App' may contain malicious libraries that expose vehicle data • More serious threats may arise when allowing access to App'
OBD-II port	• Mandatory for all car bus systems • If an infected third party device is connected, diagnostic data or malware may be installed inside the vehicle • If an intruder puts an OBD-II dongle in his car, he can continue to access sensitive information
CAN bus	• Significant weaknesses in the car interior network. Broadcast communication between ECUs connected to the CAN bus • If any ECU connected to the CAN bus is infected, it affects everyone
Wireless network	• It is possible to hack the Wi-Fi and Bluetooth networks used to connect smart phones and in-car devices
Embedded application	• Vulnerabilities in open source applications can be exploited as a means of injecting malicious software into in-car tram devices
CD player and USB port	• Vehicle connects external device to CD player and USB port • When the system is connected to the CAN bus, it acts as an interface to attack other components

wireless is described as broadcast channels, addressable channels, and stepping back [11].

According to White Paper of Charlie Miller and Chris Valasek, who are famous for white hackers in the field of vehicle security, Surfaces for remote attacks of vehicles are classified into Passive Anti-Theft System (PATS), TPMS, Remote Keyless Entry/Start (RKE) system, Telematics/Cellular/Wi-Fi, and Internet/Apps, and analyzed the remote attack potential of existing vehicles including the Audi A8 in 2014 [12] (Table 1).

4   Whitelist-Based Access Control Mechanism

In this paper, we propose a mechanism that can block illegal access when a malicious attack is made through the headunit of the vehicle through the diagnostic app of the smartphone.

The access control system is composed of a smart cluster (remote diagnosis, attack emulator), headunit, and in-vehicle network.

The smartphone accesses the vehicle's wireless AP and provides services such as door open/close, emergency light, trunk, fog light, and remote diagnosis through the diagnostic application. In the headunit, Sends a message over the bus. In the In-Vehicle Network, related ECUs process information. Since there is a limit to experiment with real vehicle, it displays information through Digital Cluster and processes the response to the message.

Android smartphone apps, GENIVI 12.0, a headunit open vehicle platform, and digital clusters developed in Linux environment (Fig. 2).

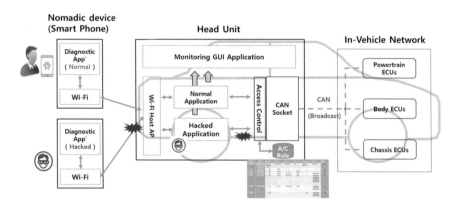

Fig. 2 Whitelist-based access control mechanism

Table 2 Example of access control-rule matrix

Subject 1	Subject 2	Resource 1	Resource 2	Resource 3	Resource 4	Resource 5	Resource 6	Rights
User ID	Application ID (name, path or Hash Value)	Protocol (TCP/UDP/CAN)	Server/Client	IP address	Port	Network device IF	Message ID	RW/M
root	/sbin/sshd	TCP	Server	*	*	*	*	M
user 001	/bin/diagnostic	TCP	Server	*	3000	*	*	RW
user 001	/bin/diagnostic	TCP	Server	192.168.1.1	3001	*	0×00000001	RW
user 001	/bin/diagnostic	TCP	Client	192.168.1.10	5000		0×00000001 0×00000002 0×00000003 0×00000005 $0 \times 0000000F$	R
user 001	/bin/diagnostic	TCP	Client	192.168.1.10	5000	*	0×00000001 0×00000002 0×00000003	W
user002	*	CAN	*	*	*	can0	0×00000001 0×00000002 0×00000003	RW

4.1 Access Control Mechanism

In this study, the threats of the vehicle were limited to the application messages of the smart phone device, the app, the headunit, and the CAN message of the in-vehicle network. Some of the hackers' attack methods are modifying and attacking smartphone apps or existing application programs of the headunit. In some cases, it is necessary to check the contents of messages because they use normal programs. We also added attack scenarios by attacking through modulated diagnostic apps and attacking with same application program as headunit's diagnostic service application.

First, add the application information of the authorized smartphone device, app, and headunit to Access Control-Rule maxtrix. It should be treated as inaccessible by general authority and, if necessary, managed by HSM. The main information to be added to the rule is Protocol (TCP/UDP/CAN), Server/Client, IP address, Port, Network device, and Message ID based on the authorized user ID and application ID.

Monitors the information of messages coming and going from the kernel of the headunit, and outputs a log to the management page window of the headunit in case of violation of the rule, and blocks the message from flowing into the in-vehicle network (Table 2).

5 Conclusions

In this paper, we have examined the connectivity and major vulnerabilities of vehicles. We then model and implement a white—list—based access control system consisting of smart phones, headunits, and digital clusters. Although it is implemented by

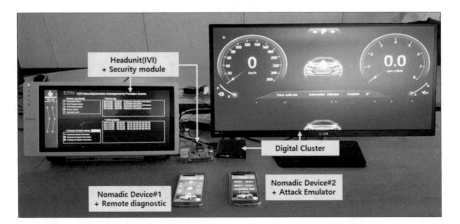

Fig. 3 Threat and access control through headunit

constructing a similar environment instead of a real vehicle, it is implemented in an open platform (GENIVI) used in a real vehicle, and designed/implemented so that it can operate even when a normal app is modulated. We plan to add an In-Vehicle Network part in order to establish a similar environment with the actual vehicle in the future. By logging, analyzing and replaying the CAN message of the actual vehicle, Dos, Fuzzy, Replay, and Impersonation attacks (Fig. 3).

Acknowledgments This work was supported by Institute for Information and communications Technology Promotion (IITP) grant funded by the Korea government (MSIT) (No. B0717-16-0097, Development of V2X Service Integrated Security Technology for Autonomous Driving Vehicle).

References

1. Kwon, H.: Security trends for autonomous driving vehicle. Electron. Telecommun. Trends **33**(1), 78–88 (2018)
2. Siemens: Vehicle-to-X (V2X) communication technology (2015)
3. WIKIPEDIA: Vehicle-to-everything. https://en.wikipedia.org/wiki/Vehicle-to-everything/
4. Lee, H., Jeong, S.H., Kim, H.K.: OTIDS: A Novel Intrusion Detection System for In-vehicle Network by using Remote Frame. PST (2017)
5. Wolf, M., Weimerskirch, A., Paar, C.: Security in automotive bus system. In: ESCAR (2004)
6. Koscher, K., Czeskis, A., Roesner, F.: Experimental security analysis of a modern automobile. SP (2010)
7. Hoppe, T., Kiltz, S., Dittmann, J.: Security Threats to Automotive CAN Networks—Practical Examples and selected Short-Term Countermeasures, pp. 235–248 (2008)
8. Wright, A.: Hacking cars. Commun. ACM **54**(11), 18 (2011)
9. Miller, C., Valasek, C.: Adventures in Automotive Networks and Control Units (2013). http://illmatics.com/
10. Coppola, R., Morisio, M.: Connected car: technologies, issues, future trends. ACM Comput. Surv. **49**(3) (2016)
11. Checkoway, S., McCoy, D., Kantor, B.: Comprehensive Experimental Analyses of Automotive Attack Surfaces. USENIX Security (2011)
12. Miller, C., Valasek, C.: A Survey of Remote Automotive Attack Surfaces. Black Hat USA (2014)

Author Index

© Springer Nature Switzerland AG 2019
R. Lee (ed.), *Computational Science/Intelligence & Applied Informatics*, Studies
in Computational Intelligence 787, https://doi.org/10.1007/978-3-319-96806-3

Printed in the United States
By Bookmasters